Explosive Welding

Explosive Welding

Processes and Structures

B. A. Greenberg • M. A. Ivanov
S. V. Kuz'min • V. I. Lysak

CISP

CRC Press
Taylor & Francis Group
Boca Raton London New York

CRC Press is an imprint of the
Taylor & Francis Group, an informa business

Translated from Russian by V.E. Riecansky

CRC Press
Taylor & Francis Group
6000 Broken Sound Parkway NW, Suite 300
Boca Raton, FL 33487-2742

Printed on acid-free paper

International Standard Book Number-13: 978-0-367-35578-4 (Hardback)

Visit the Taylor & Francis Web site at
http://www.taylorandfrancis.com

and the CRC Press Web site at
http://www.crcpress.com

Contents

1

Introduction

Thousands of different compounds are produced by explosive welding. If each of the compounds had its own formation processes, then their identification would be completely hopeless, as well as finding out the reasons for weldability. However, despite the diversity of welded joints, there are not many such processes. They can be detected and identified.

Explosive welding is an intensively developing area. The number compounds that are obtained by this method is constantly growing. Moreover, it is possible to obtain compounds of metals that have not been obtained by other methods, for example, titanium with steel, zirconium with steel and many other compounds on very large areas. It is possible to weld sheets and products of complex shape and also both dissimilar and homogeneous metals. High-quality bimetallic and multilayer composites with high strength can also be produced.

The unusual microstructure of compounds that occur during explosive welding is due to the fact that explosive welding is a high-intensity transient impact [1, 2]. Characteristic times: welding time is about 10^{-6} s, the strain rate is $10^4...10^7$ s^{-1}, the cooling rate is 10^5 K/s. The existence of the so-called 'window of weldability' (in the coordinates 'the angle of impact–the velocity of the point of contact') reflects the necessary conditions for the formation of a strong joint [3, 4]. With all the variety of materials and welding modes, the central problem is the mixing in the transition zone near the interface. It is the structure of the transition zone that determines the possibility of adhesion of both materials.

Mixing occurs as a result of a strong external impact, which implies a large plastic deformation (including pressure, shear components, stress rotation moments, non-uniform deformation, etc.), friction of surfaces, the effect of a cumulative jet and other factors.

But it still remains unclear how, even with such a strong external effect, mixing takes place in such a short time as the welding lasts. This question is even more acute if we are talking about materials that do not have mutual solubility, even in a liquid state.

A whole range of questions arise:

- is it necessary for the weldability to have atomic-clean surfaces;
- which surface – flat or wavy – is preferable for mixing;
- what role does melting play (local or along the entire interface;
- what is the role of mutual dissolution of materials;
- why immiscible suspensions are stirred;
- how is mixing is carried out when welding a metal with an intermetallic compound;
- how dangerous is the formation of intermetallic compounds when welding a metal to a metal.

When varying the welding parameters and their optimization taking into account the 'weldability window', it is necessary to study the structure of the joint and identify its main elements. However, it turned out that structural studies constitute only a small fraction of studies of welded joints. Moreover, the role of structural research is still undervalued.

As a result, a discrepancy has arisen: some stereotypes, previously taken on trust, are not confirmed by detailed structural studies. We are talking about such generally accepted views as the need for atomically clean and atomically smooth surfaces, the need to form active centres on approaching surfaces, the danger of melting, the risk of lack of mutual solubility, the existence of only one type of fragmentation inherent in intense plastic deformation.

On the basis of experimental data, a number of hypotheses on the nature and mechanisms of interaction of metals in the solid phase have been proposed [5, 7]. However, these hypotheses are not consistent with each other and often contradict the results of experimental studies.

The film hypothesis is based on the assumption that the formation of a compound requires the convergence of the clean (juvenile) surfaces of the joined metals to the interatomic distance. The compound is formed as a result of grasping, which is a diffusion-free process of combining crystal lattices of the jointly deformed metals.

The recrystallization hypothesis is based on the assumption that the mechanism for combining contacting metals whose crystal lattice is distorted lies in the recrystallization process, which results in the formation of new crystals common to the metals being welded.

The diffusion hypothesis suggests that a joint is formed as a result of the mutual diffusion of atoms of the contacting surfaces. Getting a strong joint is explained by the occurrence of normal metallic bonds as a result of local deformation at elevated temperature, the convergence of surfaces, the maximum increase in area and mutual diffusion of atoms of the metals being joined. The ability to join mutually insoluble metals, observed in practice, is explained by the possibility of mutual diffusion in a very thin layer of those metals that are considered insoluble in each other.

According to the energy hypothesis, the convergence of pure metal surfaces with differently oriented crystals over the interatomic interaction distance is a necessary but still insufficient condition for the formation of a welded joint. For the formation of metallic bonds, it is necessary that the energy of the atoms of at least one of the surfaces to be joined exceeds a certain level characteristic of this metal: the energy threshold of grasping. Overcoming the energy threshold of grasping is due to the need to match the direction of the bonds or the transition of the metal in an amorphous state.

According to the vacancy (dislocation) model, the process of plastic deformation is accompanied by the displacement of surface masses in the contact zone to a depth of tens of microns, causing displacement of point defects. It is believed that the joining of metals can be carried out under the condition of successive processes: convergence of surfaces at an interatomic interaction distance, as well as an increase in the density of point defects (vacancies and dislocated atoms) in the contact field and, finally, the formation of a joint due to mass transfer during displacement of point defects.

The physical basis for the joining and weldability of metals during pressure welding is the natural metallic bond. The pressure applied in the process of cold welding causes directional deformation, with the help of which the surfaces are cleaned and results in the directionality of bonds, the convergence of atoms, as well as the increase of their energy levels to the state necessary for the appearance of metal bonds.

The hypothesis of topochemical reactions (active centres) gives a quantitative dependence of the strength of the welded joint on the physico-chemical and mechanical properties of the metals being joined. Based on the general theory of the imperfections of the crystal lattice and the kinetics of chemical reactions, it is assumed that the grasping of metals is considered as a special case of topochemical reactions during pressure welding, which are characterized by a three-stage process of forming strong bonds between the atoms of the metals being joined: activation of contact surfaces; volume development of interaction.

Without dwelling on the analysis of these models, we note only the following. Due to the high speed of explosive welding, the use of thermally activated processes in welding models requires caution. First of all, it is diffusion, as well as such processes of dislocation movement as cross slip and climb. At the same time, the generation of dislocations under the action of external stresses occurs without thermal fluctuations and is certainly possible. The occurrence of thermally activated processes apparently becomes possible only at residual temperatures. This is a significant difference between explosive welding and diffusion welding, where thermally activated processes are dominant. A comparison of these types of welding was carried out in [8, 9] for joints of titanium with orthorhombic titanium aluminide.

Outwardly simple, but in its physical essence very complex, the process of explosive welding is a little similar to other methods of joining materials and requires not only detailed structural analysis, but also a new approach, which includes previously unexplored processes and their relationship to related processes (see the reviews [10...14]).

The fastest of the processes of formation of welded joints, which occurs even during the explosion, is the scattering of particles. We believe that the scattering of fragments in an explosion and the scattering of particles during explosive welding have much in common. However, in an explosion, fragments fly apart in open space, while in explosive welding, particles fly apart in the closed space between the plates. When welding, the role of an explosion is to disperse one body relative to another, so that a large part of the chemical energy of an explosive goes into the kinetic energy of the flyer plate. It is the kinetic energy that passes into other types of energy upon impact [1...3].

The authors of the monograph discovered a fragmentation of a new type, which was called granulating fragmentation (GF) [12...14]. Although the explosion and explosive welding are accompanied by the formation and dispersal of particles, these types of fragmentation are not identical. GF is a process of separation into particles, which either scatter or merge with each other. In other words, the GF includes both the scattering of particles and their partial consolidation.

GF is an analogue of fragmentation in an explosion, but occurring in the presence of various strong barriers. For particles (fragments), emitted during explosive welding from a single plate, such obstacles that stop them will serve both the second plate and the main mass of the original plate. In this case, it can be assumed that the expansion of solid particles of a phase that does not undergo melting will initiate a local melting of an easily fusible material near the interface. This is due to the fact that, due to the large total area of the particles, the effective friction between the particles and the barrier can cause local heating, sufficient for melting. This is confirmed by the observation of numerous particles of the refractory phase inside the local melting zones. Local melting zones are filled with either a true solution or, for immiscible phases, a collodion solution.

The chain of processes occurring in GF – explosion, the formation of particles (fragments), local dispersion of particles, local heating of the obstacle, local melting – make melting during explosive welding almost inevitable. This is an unexpected (and for the authors themselves) result that requires discussion and expansion of the field of research. All studied joints showed melting (either local or along the entire interface).

The presence of melting zones can be one of the most essential mechanisms for the adhesion of materials during explosive welding. Indeed, for many materials, the best bonding substance is the solution or melt of this substance. When melting, problems of wetting, adhesion, thermal expansion and protection against contact corrosion are solved immediately. However, such a positive effect of the growth of adhesion takes place far from always, but depends on the ratios between the characteristic temperatures of the contacting materials and is determined by the conditions for the formation of various types of colloidal solutions. In particular, it is shown that, depending on the external conditions, the melting either provides bonding or leads to the formation of risk zones.

In addition to fragmentation and melting, the process controlling the formation of a welded joint is the formation of cusps on the interface. We believe that the projections are the 'wedges' that ensure the adhesion of the contacting surfaces. An unusual form of cusps has been found: although they are solid-phase, they look like splashes on water. The evolution of the interface with the intensification of the welding regime includes a number of transition states containing both waves and splashes. We can assume that in these cases the interface has a quasi-wave shape. Previously, transient states were not observed.

The proposed approach allows us to understand and consistently explain the set of the obtained experimental facts. The sequence of sections corresponds to the internal logic of the proposed approach: after Chapter 1 (Materials and joints), Chapter 2 presents the experimental results of studying the structure of various joints. Chapter 3 discusses the formation of welded joints during explosive welding. In addition to explosive welding, another type of severe plastic deformation (SPD) is studied, namely SPD torsion. The evolution of the structure is studied during the consolidation of ceramic powders (quartz, crystal) and glasses (object, quartz). Chapter 4 begins with the fastest process (section 4.1). This is the granulating fragmentation (GF), an analogue of fragmentation during the explosion. It is the GF that causes local meltdown to which section 4.5 is devoted. In addition to local melting zones, inhomogeneities of another type of interface are investigated, namely cusps (section 4.4). The processes discussed in Chapter 4 may in some cases contribute to weldability, in other cases, on the contrary, lead to the formation of risk zones (Chapter 5). Obtaining a three-layer composite for the wall of a chemical reactor with high stability under harsh conditions of long-term operation is one of the most successful implementations of explosive welding (section 5.1). In Chapter 6, taking into account the self-similarity of splashes, as well as particles inside the local melting zones, their fractal dimensions are calculated. The following chapters are devoted to the study of the evolution of the interface for copper–tantalum, aluminum–tantalum (Chapter 7), copper–titanium (Chapter 8), copper–copper (Chapter 9). Chapter 10 covers issues related to the production and study of the microstructure of multilayer composites, their comparison with two-layer composites, as well as with composites obtained by SPD torsion. Self-organization processes that determine transitions between splashes, large groups of splashes and a quasi-wave interface (Chapter 11) have been studied for the first time.

The authors of the monograph for the first time discovered the following processes and structures: granulating fragmentation, cusps, splashes, and a quasi-wave interface.

2

Materials and joints

Metal–intermetallic welded joints were investigated. Technically pure titanium was chosen as the metal, and titanium aluminide was used as the intermetallic compound (hereinafter referred to as aluminide for short). Depending on the aluminide composition and the welding conditions, various compounds were obtained, which for convenience were named as follows: ($\mathbf{A}_w$), ($\mathbf{A}_p$), ($\mathbf{B}_w$), ($\mathbf{B}_p$). Here the subscript indicates the shape of the boundary (planar–wavy). VTI-1 orthorhombic alloy, containing 16 at.% Nb, was used for compounds **A**, and VTI-4 orthorhombic alloy containing 23.5 at.% Nb was used for welded joints **B**. For some joints, melting along the entire boundary was observed; for others, local melting zones with a vortex structure were detected.

Metal–metal welded joints were also investigated. In order to find out how important the presence of mutual solubility of starting materials was, metals (copper–tantalum, iron–silver) were chosen for explosive welding; aluminum–tantalum, steel–steel, copper– titanium, copper–copper pairs, which have mutual solubility, were also studied.

Explosive welding experiments were carried out by the CRISM 'Prometey' (St.Petersburg), Volgograd State Technical University, Ural Chemical Engineering Plant (Ekaterinburg). Welding was performed with different procedures and parameters, after which the welds were selected for further study, which, for convenience, are indicated here as follows:

($\mathbf{A}_w$) – titanium – VTI-1, the wavy boundary;

($\mathbf{B}_w$) – titanium – VTI-1, flat boundary along which melting took place;

($\mathbf{B}_w$) – titanium – VTI-4, wavy boundary;
($\mathbf{B}_p$) – titanium – VTI-4, almost flat, partially melted boundary;
($\mathbf{C}_p$) – copper–tantalum, flat boundary;
($\mathbf{C}_w$) – copper–tantalum, wave-like boundary;
($\mathbf{D}_w$) – iron–silver, wavy boundary;
($\mathbf{E}_p$) – aluminum–tantalum, flat boundary;
($\mathbf{E}_w$) – aluminum–tantalum, wavy boundary;
($\mathbf{S}_w$) – steel–steel, wavy boundary.

A parallel study of compounds with both flat and wavy boundaries for the same pair of dissimilar materials is extremely rare. However, as can be seen from further analysis, such a comparison turned out to be quite successful.

Figure 1 gives the welding parameters for all the studied joints: γ is the angle of impact; V_c is the speed of the contact point. Note that the same modes were used to obtain two joints ($\mathbf{A}_p$) and ($\mathbf{B}_w$).

In addition, to study the transition states of the interface, intermediate modes were used; the parameters of these modes for copper–tantalum and aluminum–tantalum welds are given in chapter 7, and for copper–titanium welds in chapter 8.

Experimental studies of the microstructure were performed using the following methods: X-ray diffractometry (DRON 3), scanning (SEM) and transmission (TEM) electron microscopy: JEM200CX and SM-30 Super Twin electron microscopes, QUANTA 200 FEI Company. In the study of the evolution of the microstructure during

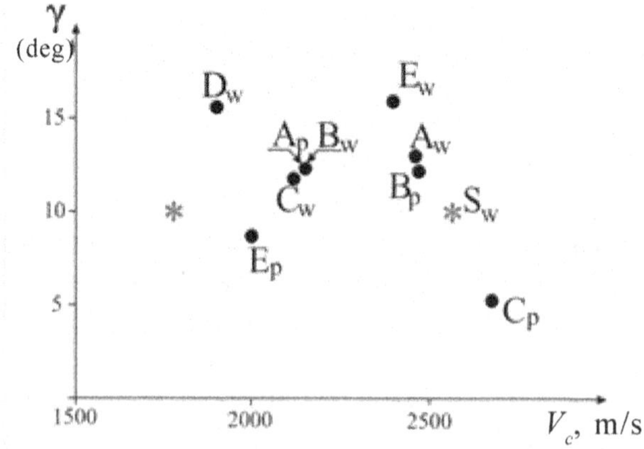

Fig. 1. Investigated compounds: γ – angle of impact; V_c – the point of contact velocity.

the heating process an attachment to the scanning microscope was used. The Zygo NewView 7300 optical profilometer was used to study the surface roughness of the starting materials. The equipment of the Center for Diagnosing Structure and Properties of Materials (Belgorod) was also used: an OLYMPUS GX51 optical inverted microscope; scanning electron microscope Quanta 600; ion cannon Fashione 1010 ION MILL. Microhardness was measured by means of a PMT-3 microhardness tester.

The reason for the insufficient extent of electron microscopic studies of welded joints is the difficulty of making foils from dissimilar materials. In the process of thinning one of the metals can be completely etched, so that the foil remains only from another material. Thus, for a copper–tantalum joint with high corrosion resistance of tantalum, only a special selection of reagents makes it possible to prevent copper etching. The most effective is the use of an ion cannon.

When using the above-mentioned intermediate modes, special attention is paid to transitions through the lower boundary (LB) of the 'weldability window' and clarification of the reasons why weldability is impossible below LB, although grasping is possible. The lower boundary of the 'weldability window' is important both for practical calculations of welding conditions and for understanding the processes that determine the possibility of the formation of a welded joint. Welding modes near LB are characterized by minimal impact speeds, ensuring the formation of a strong joint. As noted in [3, 4], most authors of publications believe that the position of LB depends on the strength characteristics of materials, such as hardness, tensile strength or yield stress. According to R.H. Whitman (see [3]), the LB position is determined by the critical impact pressure, providing plastic flow near the interface, and is calculated through the minimum impact speed required for welding $V_{c\min} = \sqrt{\frac{\sigma_B}{\rho}}$,where σ_B is the tensile strength, ρ is the density. In [3, 4], specific expressions for $V_{c\min}$ obtained in different models are given. In these expressions, the root dependence of $V_{c\min}$ on some specific stress is preserved, for which either the tensile strength or yield stress or hardness is used. In addition, some of these expressions take into account the parameters of heat exchange processes, viscosity coefficients, etc. At the same time, it remains unclear which of the parameters of the two dissimilar metals are decisive.

In addition to the 'force' approach of R.H. Whitman, with which the models mentioned above are related in one form or another, there is another 'energy' approach based on the so-called Astrov criterion (see [3]). In this case, the position of the LB is determined by the constant and characteristic for each concrete pair critical energy W_{2cr} spent on plastic deformation. When welding dissimilar materials, this value is close to the energy required for plastic deformation of a softer material. For the joints under study, the low-melting phases are copper and aluminium. The question of the theoretical determination of the LB position and the determination of the corresponding critical values $V_{c\ min}$ and W_{2cr} still remains largely open. The research conducted in this work is not aimed at determining the specified critical values, but at identifying the characteristic structure of the interface, which must be achieved in order for welding to occur.

3

Experimental results

3.1. Titanium–orthorhombic titanium aluminide

The Ti–Al–Nb system alloys form a large group of alloys [15, 16], including, on the one hand, those that have a reduced Nb content and are based mainly on the α_2-phase (HCP), and on the other hand, enriched in Nb and based on the orthorhombic O-phase (Fig. 2). The O-phase is actually a slightly distorted form of the α_2-phase. In addition to the indicated phases, depending on the composition, orthorhombic alloys may also contain BCC phases: a disordered β-phase or an ordered β_0 (B2)-phase.

Consider a section of an isothermal (T = 1173 K) section of the phase diagram of the Ti–Nb–Al ternary system (Fig. 3) [17]. Here, there are areas of existence of the alloy in single-phase α-,

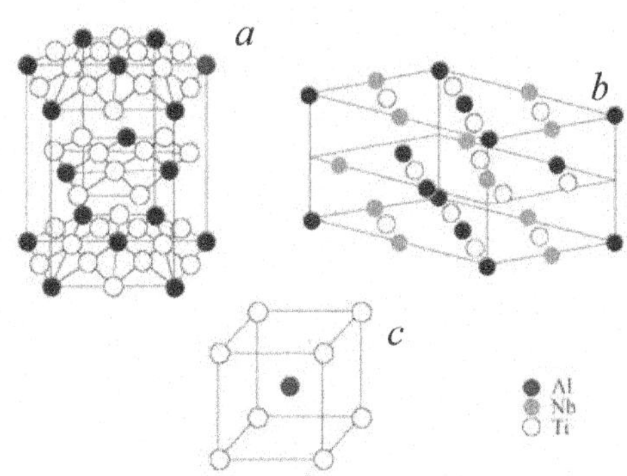

Fig. 2. The elementary cells of the phases of orthorhombic titanium aluminide: a – α_2, b – O, c – B2 phases.

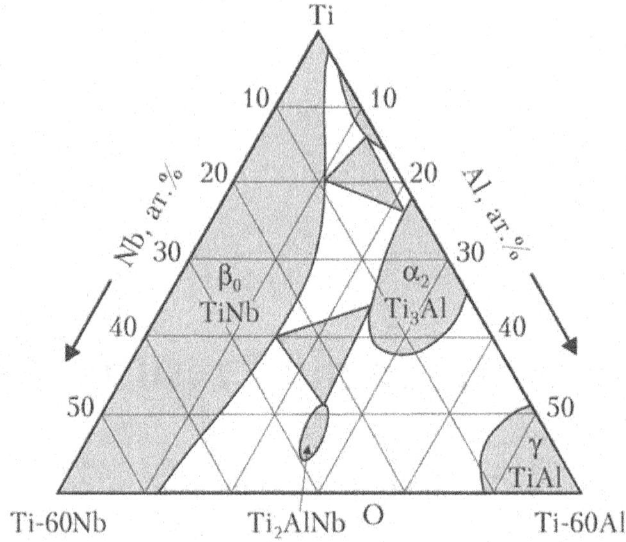

Fig. 3. Isothermal section (T = 1173 K) for the Ti–Nb–Al ternary system.

Table 1. Explosive welding parameters

Compound	β	V_{con}	V_c
(**A**$_w$)	13	2200	500
(**A**$_p$)	12.3	2150	460
(**B**$_w$)	12.3	2150	460
(**B**$_p$)	12.18	2469	524

β-, α_0-, α_2-, γ-, or O-states, as well as areas of existence of three-phase (α–α_2–β)- and (O–α_2– β_0)-alloys. The latter is the required region of existence of promising three-phase alloys based on the orthorhombic intermetallic compound with a strong deviation from the stoichiometric composition.

The parameters of explosive welding (γ – impact angle (deg); V_{con} – contact point velocity (m/s), V_c – impact velocity (m/s)) of VTI-1 and VTI-4 orthorhombic alloys with technically pure titanium VT1-0 are given in Table 1.

Orthorhombic titanium aluminides were chosen because of their inherent complex strength properties, including high values of specific strength (strength/density ratio), good plasticity at room temperature, fracture toughness, high creep and oxidation resistance [18, 19]. For orthorhombic aluminides of certain compositions, it is also possible to obtain good tensile properties, which is unusual for

TiAl and Ti_3Al [20, 21]. The excellent combination of the strength properties inherent in the orthorhombic titanium aluminides gives reason to hope for their successful use in designing bimetallic joints. Particularly attractive is the possibility of implementing as one of the phases a disordered solid solution, which, with its increased ductility and toughness, can serve as a damping element in the structure of a multiphase alloy based on intermetallic compounds.

Analysis of the structure of compounds for different compositions of the orthorhombic alloy and a different shape of the interface, was carried out by the authors in [22–26] by diffractometry, TEM and SEM. When deciphering the phases and structures, the data presented in [27–29] was used.

3.1.1. (A_w): Titanium – VTI-1, wavy boundary

The following thermomechanical treatment was used for the VTI-1 orthorhombic alloy (Ti–30Al–16Nb–1 Zr–1 Mo), the composition of the alloy is indicated in at. %. A sheet of VTI-1 alloy (1.2 mm thick) was prepared and thermally processed according to the technology [30], including multi-stage processing, the last annealing was carried out at 700°C for 3 hours. The mechanical properties of the aluminide in the initial state: along the sheet $\sigma_{0.2}$ = 1150 MPa, $\sigma_в$ = 1290 MPa, δ = 4.1%; across the sheet $\sigma_{0.2}$ = 1114 MPa, $\sigma_в$ = 1291 MPa, δ = 6.6% ($\sigma_{0.2}$ – yield stress, $\sigma_в$ – tensile strength, δ – ductility). Titanium ingots were subjected to hot deformation and subsequent annealing at 750°C for 2 h. As a result, titanium had a perfect structure with an average grain size of about 50 μm.

Figure 4 *a, b* is a diagram of explosive welding. A recess was made in the lower titanium plate (substrate) into which a sheet of VTI-1 alloy was laid and fixed. An explosive charge (EC) was placed on the top (cladding) titanium plate.

During the explosion the velocity of the plate reached 500 m/s, and the angle of impact was 12...14°. At explosive detonation on the contact surface (CS), a pressure of ~6 GPa developed, and the material adjacent to the CS was heated to ~900°C and was subjected to plastic deformation by 40...80%. Figure 4 shows schematically the weldability window.

The diffractograms of the starting materials are shown in Fig. 5 *a* and *b* respectively. The aluminide of the studied composition contains a hexagonal α_2-phase (structure $D0_{19}$) and an orthorhombic

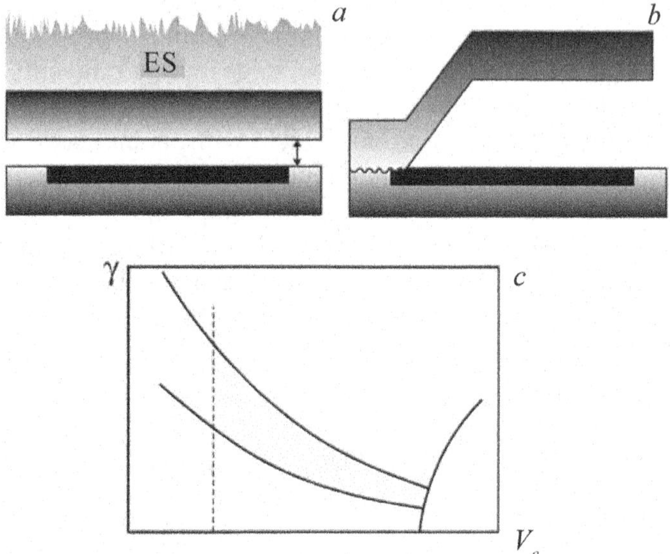

Fig. 4. Scheme for obtaining bimetallic joints by explosive welding (*a, b*); image of the 'weldability window' (*c*).

O-phase. In addition, there is a small amount of BCC-phase with traces of ordering according to type B2.

Optical micrographs (Fig. 6) show that the structure of·the VTI-1 orthorhombic alloy represents the distribution of regions of a globular shape against the background of a homogeneous structural component. Large (~20 μm) globules are clearly visible, between which there are small (~2 μm) globules. As shown by the TEM study, the globular regions have a layered internal structure and represent the α_2-phase, which is at different stages of decay. Next, the globular structure of the aluminide will be used as a witness of the transformations that occur during the explosion.

The diffraction patterns of the titanium aluminide after welding are shown in Fig. 5 *c*. As can be seen from the comparison of Figa. 5 *c* and 5 *b*, far from the contact surface after welding the aluminide retains almost completely set of X-ray peaks observed in the initial state (O + α_2), but the lines identified as B2 disappeared. There was a slight decrease in the intensity and broadening of X-ray lines, indicating the work hardening of the initial material. Thus, during explosive welding, in the main volume of the aluminide residual metastable phases with a BCC lattice were suppressed due to their transformation into the orthorhombic (O) and hexagonal (α_2) phases.

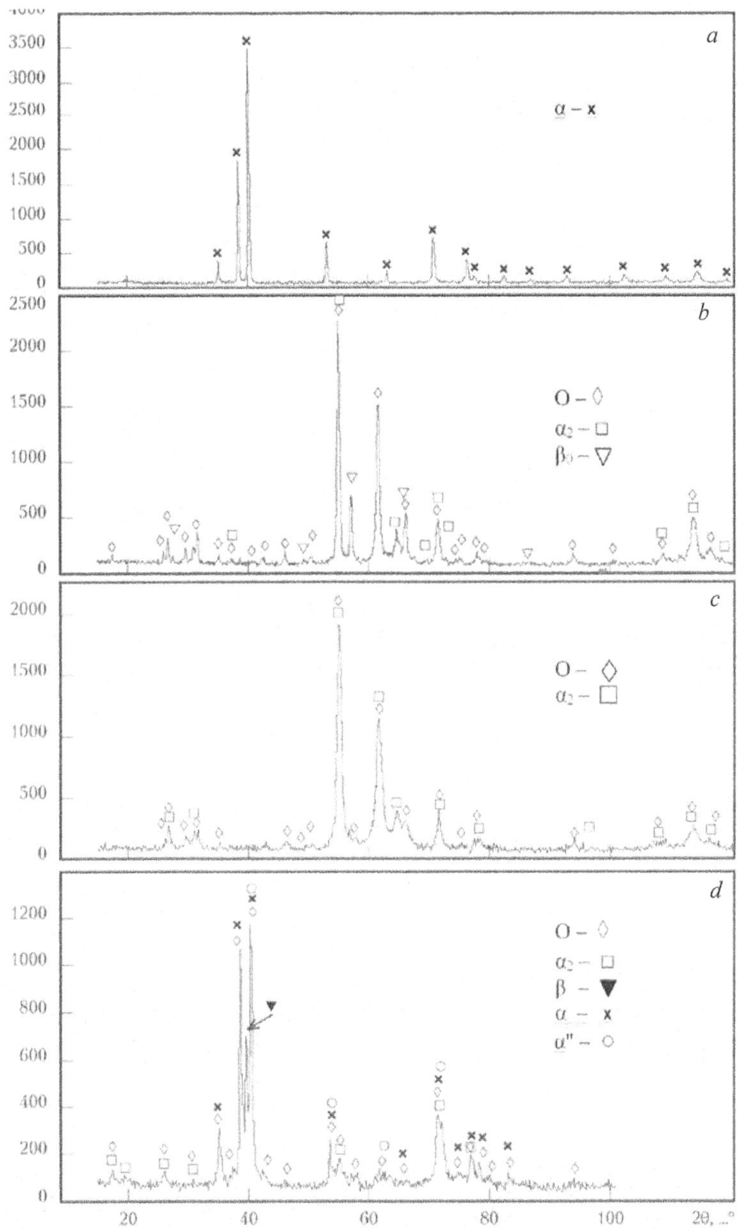

Fig. 5. Diffractograms for the initial materials and for the compound ($\mathbf{A}_w$): *a* – α-Ti in the initial state; *b* – VTI-1 orthorhombic alloy in the initial state; *c* — VTI-1 orthorhombic alloy after deformation by an explosion; *d* – weld area.

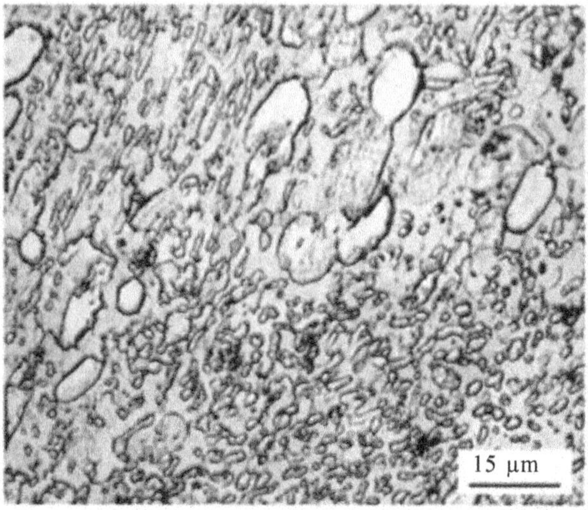

Fig. 6. The structure of the VTI-1 orthorhombic alloy in the initial state.

Only the hexagonal α-phase was detected in a titanium cladding plate at different distances from the contact surface. The diffractogram differs from the initial one only in some blurring of the peaks.

The diffraction patterns of the heat-affected zone were taken from the section which, due to the waviness of the contact surface, contained areas of the orthorhombic alloy and the cladding plate of commercially pure titanium. For this reason, the diffractogram of the heat-affected region (Fig. 5 *d*) contains peaks belonging to both aluminide and α-titanium. Peaks have a strong blur, as a result of which some of them can be interpreted as a set of reflections from different phases with close reflection angles. As for the BCC phases, the diffractogram (Fig. 5 *d*) shows only a peak at $2\theta = 39.5°$ which

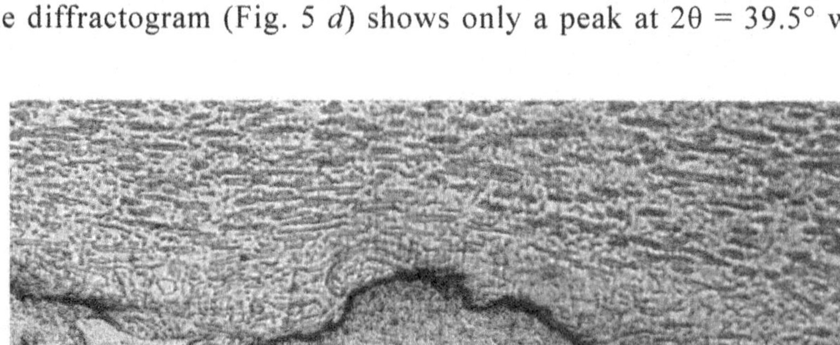

Fig. 7. The cross-section of the wavy interface for the joint ($\mathbf{A}_w$): *a* – the upper band – VTI-1 orthorhombic alloy with globular regions; lower band – titanium (optical micrograph).

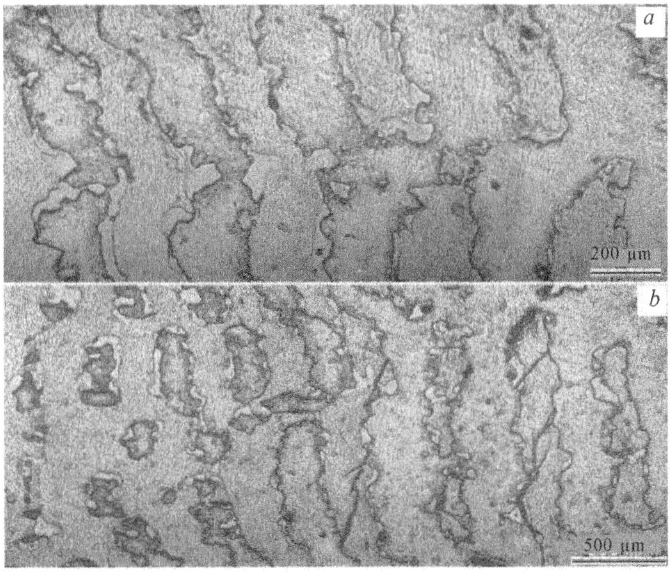

Fig. 8. Weld ($\mathbf{A}_w$), longitudinal section (optical micrograph): alternating bands of aluminide and titanium at different magnifications.

can be attributed to them. However, it is not possible to identiy other peaks.

The optical micrograph (Fig. 7) for the weld under study ($\mathbf{A}_w$) shows the cross-section of the interface having a wavy (corrugated) shape. The initial metals are easily distinguishable, thanks to the highly visible globules in the aluminide. The wavelength and amplitude are approximately 300 μm and 30 μm, respectively. The isolated regions are visible, which, as will be shown below, are melt zones.

The optical micrograph (OM) (Fig. 8) shows a longitudinal section of the interface. One can see the alternation of the bands of the starting metals (the light band is titanium). It is also clearly seen that the corrugated interface is not smooth, but contains cusps of one material in another, which are tens of microns. Local melting zones are also visible here.

Figure 9 shows large-scale incomplete turns (macroturns) of the material. The image of rotations of the material that became visible due to the globules playing the role of natural markers is actually a microphotograph of the rotational mode of plastic deformation. The formation of macroturns occurs in a solid and is not associated with melting.

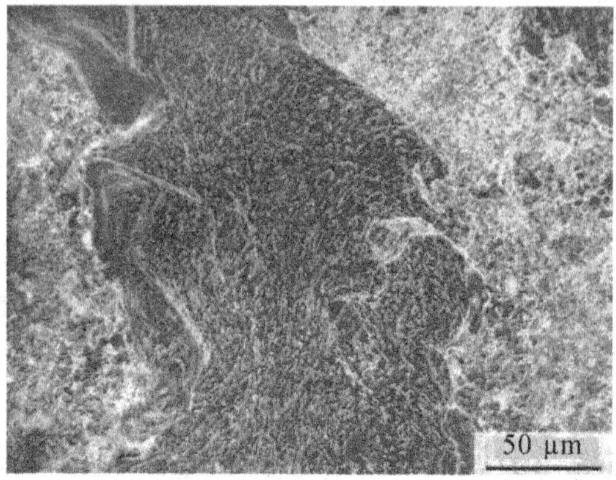

Fig. 9. Compound (A_w), longitudinal section: macroturns in titanium aluminide (SEM).

SEM photomicrographs in Fig. 10 demonstrate the first observations of cusps on the interface [23]. The depth of penetration of the cusps into another material is tens of microns. The cusps were first discovered by the authors of the book, who suggested the name of this defect. The cusps were observed in all the compounds investigated in the present work. Initially, we assumed that the cusps have a shape close to conical.

Indeed, this form have the cusps on the wave-like border in various welded joints. But the shape of the cusps on a flat border is different: they are similar to 'splashes' (see section 4.4). Further, their role in the adhesion of the contacting surfaces is discussed in detail. Moreover, their particular role may be that they are precursors of a wave-shaped interface.

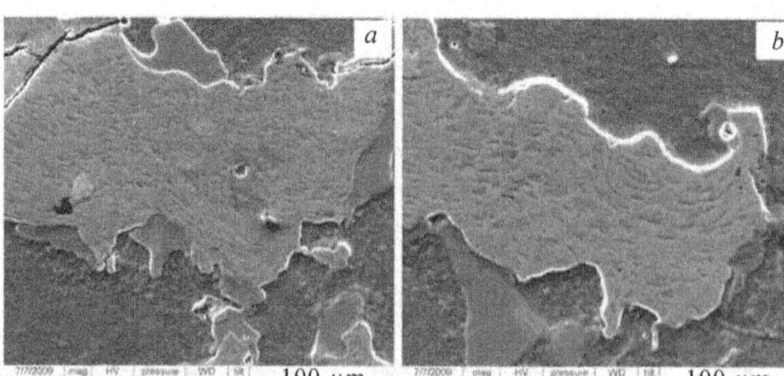

Fig. 10. Welded joint (A_w), longitudinal section of wavy interface (SEM), cusps; band with a globule – aluminium.

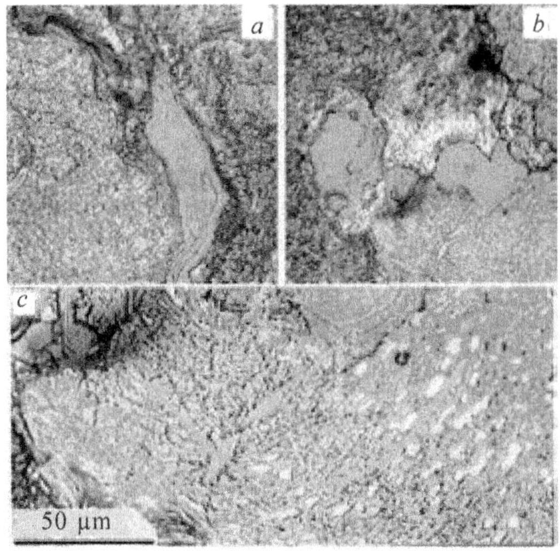

Fig. 11. Welded joint ($\mathbf{A}_w$), cross sections (optical micrograph): *a, b* – structural elements of the transition zone, highlighted with colour: yellow colour – aluminide; blue – titanium; pink – recrystallized zone of titanium; green – whirlwinds; *c* – macroturn.

Due to the presence of the globules, Fig. 10 shows the macroturns of titanium aluminide, similar to those seen in Fig. 9. The various structural elements of the transition zone are visible in Fig. 11, where the cross-section of the wavy interface is shown. Macroturns are clearly visible in the aluminide. Local melting zones with a vortex structure are clearly visible. The recrystallized zone of titanium in the form of a narrow strip of grains is clearly visible. The grain sizes vary within 1...5 μm, i.e. it is an order of magnitude smaller than the original titanium grains (50 μm). It can be assumed that fine titanium grains compared with the initial ones result from dynamic recrystallization occurring in a strongly deformed region. The recrystallization zone was not observed in the aluminide. The lack of recrystallization of the aluminide may be due to the insufficiently high temperature and short heating time: due to the low thermal diffusivity and thermal conductivity of the aluminide, the bulk of heat is removed from the impact zone through titanium, where recrystallization is observed.

It was found that, among the many unusual structures, there are, on the contrary, structures similar to those that are observed in highly deformed solid materials. Figure 12 shows the results of the observation of a highly deformed structure obtained for the welded joint ($\mathbf{A}_w$). The thickness of the sheet of VTI-1 alloy was 1.2 mm.

The TEM image in Fig. 12 was obtained outside the heat-affected region, at a distance of approximately 0.3 mm from the interface. Figure 12 *a* is a panoramic snapshot of the structure of a rather long (~10 μm length) section of the aluminide. Here one can see the main elements of the deformed structure: zones filled with the O-phase, containing plates (strips) of different directions and sizes; colonies of individual grains of O and α_2-phases with the size of 3...5 μm.

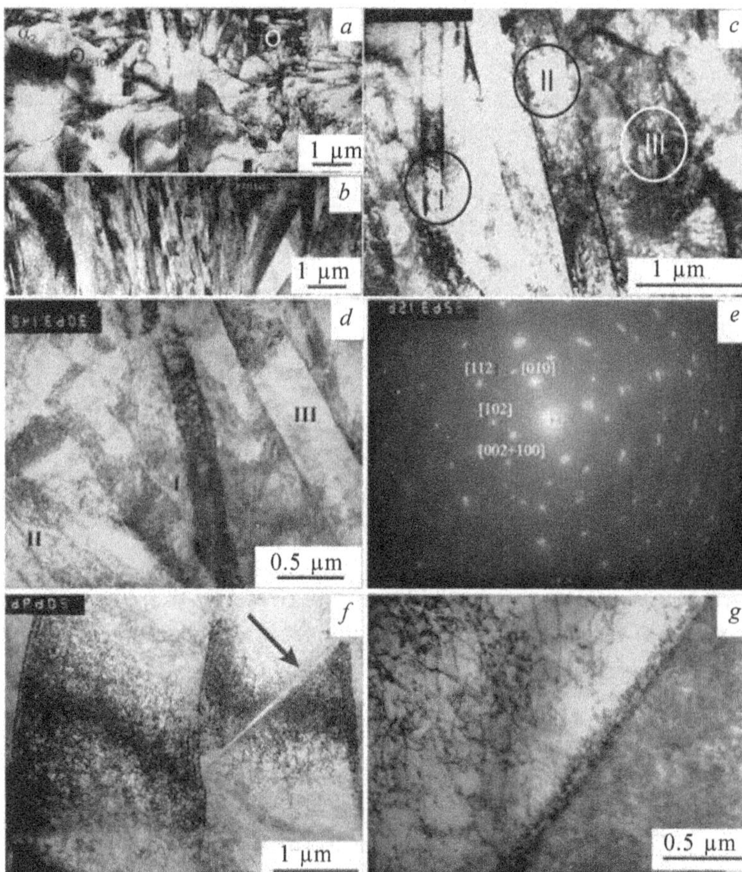

Fig. 12. Typical microstructures for the welded joint ($\mathbf{A}_w$) after explosive welding in titanium aluminide (*a...c*) and titanium (*d...g*): *a* – zones of different morphology of O- and α_2-phases; *b* – banded structure of the deformed O-phase; *c* – a site of the O-phase with a strip structure; *d* – strip structures at a depth of 2.5 mm from the outer surface of the cladding plate; *e* – electron diffraction pattern from the site shown in Fig. 12 *d*; *f* is the interaction of curved mechanical twins; *g* – the early stages of the formation of a mechanical twin at a depth of 0.5 mm from the outer surface of the cladding plate.

The high dislocation density is characteristic of all phases regardless of morphology. In the middle of the panoramic image shown in Fig. 12 *b* the region of a weakly fragmented structure is clearly visible. To the right of it, one can observe various variants of strip (dipole) configurations formed by the boundaries of deformation origin. In the dipole microstructure shown in Fig. 12 *c*, electron diffraction patterns from neighbouring sites showed misorientations on several boundaries of deformation origin. Analysis showed that the regions I and III were misoriented relative to the surrounding volumes by ~6°, and region II by approximately 12°. Figure 12 *d–e* shows characteristic types of microstructures of commercially pure titanium after explosive welding. The thickness of the titanium sheet was about 4 mm. Figure 12 *d* shows a section located at a distance of 2.5 mm below the surface in contact with the explosive, and Fig. 12 *f* is a section located at a distance of 0.5 mm from it. In addition, Fig. 12 *e* presents the electron diffraction pattern of the site shown in Fig. 12 *d*. On the example of these micrographs, it can be seen that when in explosive welding commercially pure titanium typical strip structures develop far from the boundary. Fragmentation is realized

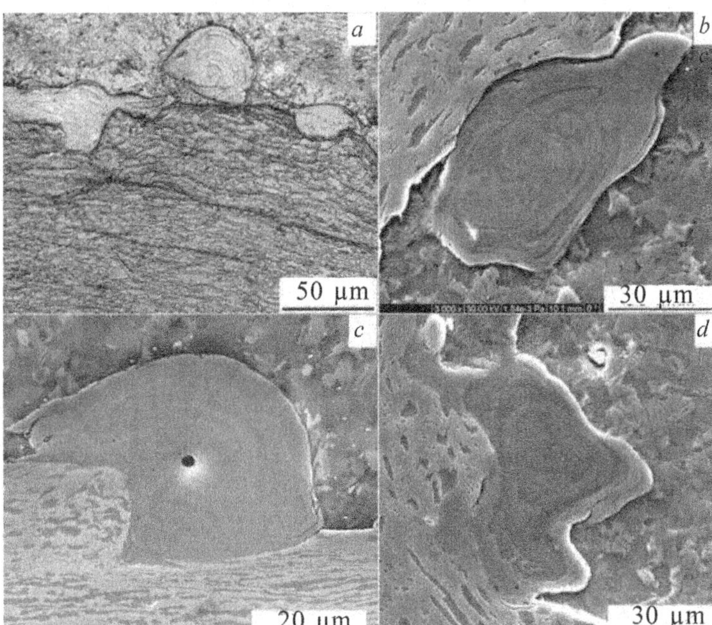

Fig. 13. Compounds (A_w), localized melt zone near wavy interface (VTI-1-titanium): a – optical micrograph; *b...d* – layered zone structure (SEM).

against the background of a high density of uniformly distributed lattice dislocations by means of mechanical twins (Fig. 12 *d*), mutual complex intersection with each other (Fig. 12 *f*). An assessment of characteristic misorientations at the boundaries of deformation origin can be made by blurring and splitting reflections on the electron diffraction pattern (Fig. 12 *e*). With acceptable accuracy, the maximum misorientation angle can be considered equal to 10°.

Localized melt zones are observed near the wavy interface. They have a layered structure in the form of concentric rings. Some typical examples are shown in Fig. 13. As can be seen from Fig. 13 *a*, the layers are already visible on the optical micrograph. The vortices are located near the weld and are clamped between the aluminide and titanium. In most cases, the main part of the vortex zone is in titanium, so that the geometric centre of the zone is approximately 15–20 μm from the interface (Fig. 13 *b*...*d*). The outer surface of the zone in some cases is a simple convex surface, but in most cases it is a surface of complex shape. Rings in shape follow the contour of the zone. As a rule, the rings are not observed near the centre of the local zone (Fig. 13 *b*, *d*). Sometimes there is a void in the centre (Fig. 13 *c*) possibly caused by shrinkage due to the difference in the specific volumes of the liquid and the solid. On the SEM micrographs fine grains are clearly visible near the vortex zone and form a recrystallized zone of titanium (Fig. 13 *b*...*d*).

Data on the chemical composition of the vortex zone obtained from with the help of SEM according to numerous measurements, which are made for the vortex shown in Fig. 14 a. The measurement zone is approximately 300 nm. In Fig. 14 *b* to the left of the

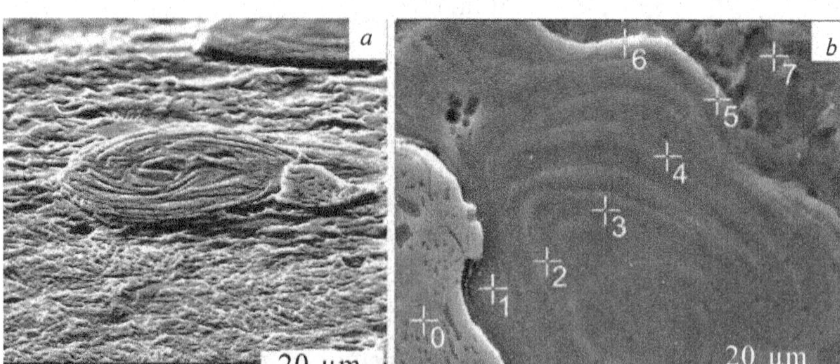

Fig. 14. Weld (A_w), SEM images of the vortex zone in secondary electron backscattering (EBSD): *a* – 70° tilt; *b* – zones of chemical composition measurement.

Table 2. Results of point measurements of the chemical composition in the vortex zone (VTI-1 titanium)

Measurement No.	Concentration, at.%		
	Al	Nb	Ti
0	25.72	17.28	57.00
1	10.71	6.77	82.52
2	11.21	7.15	81.64
3	10.63	7.43	81.94
4	11.04	7.27	81.69
5	07.21	4.27	88.51
6	07.36	4.13	88.51
7	1.2	0.64	98.27

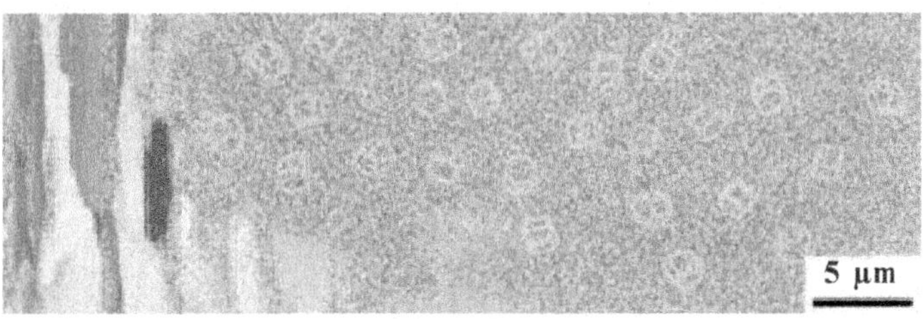

Fig. 15. Weld (A_w), SEM image of the central part of the vortex zone (EBSD method, secondary electron backscattering).

aluminide vortex zone (globules are visible on a white background), to the right – the recrystallized zone of titanium. Table 2 shows the concentrations of elements. As can be seen from Table 2, the chemical composition differs from the composition of the starting metals: aluminide (Ti–30 Al–16 Nb for basic elements) and titanium. The observed composition of the zone indicates the penetration of elements from one material into another. In fact, the main part of the vortex zone contains a solid solution based on titanium (≥80% Ti).

In the micrograph (Fig. 15) obtained using the EBSD method (backscattering of secondary electrons), it is clear that the central part of the vortex consists of phases with hexagonal and cubic lattices (see the above color insert).

The optical and SEM microscopy data allow, to some extent, to judge what constitutes a transition zone. Near the interface, which has a wavy shape, the transition zone contains isolated vortex regions, but there are parts of the interface, where the transition zone is almost invisible; in its composition, the transition zone is a set of solid solutions of different compositions, different from the initial ones. But it remains unclear which phases form the transition zone and whether the ordered phases remain.

The micrograph (Fig. 16 *a*) shows a panorama from a long section of the transition zone. In the micrograph (Fig. 16 *b*), obtained in the (002)α reflex, luminous α-phase grains are seen against the α-phase grains with different orientations and β-phase grains. The microdiffractogram (Fig. 16 *c*) shows reflections from several orientations (reciprocal lattice sections) of the α-phase and reflexes of the β-phase zoe axis [331] (with diffuse strands). In the micrograph (Fig. 16 *d*) obtained in the (110) reflex, the β-phase grain glows (see microdiffraction in Fig. 16 *e*).

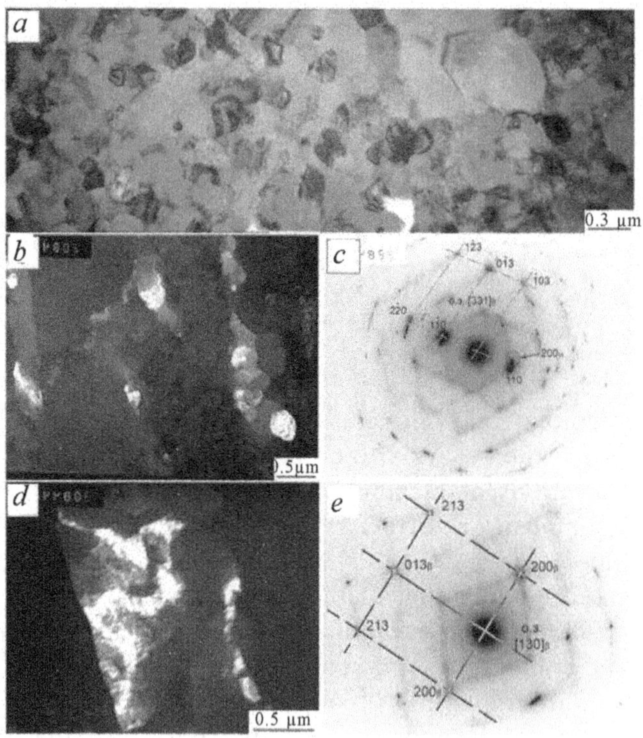

Fig. 16. Welded joint (A_w), the microstructure of the transition zone: *a* – panorama of the extended part of the vortex zone; *b* – grains of the α-phase in the (002) α reflection (dark field); *c* – microdiffraction from thefigure shown in Fig. 16 *b*, reflexes from several orientations (cross-sections of the reciprocal lattice) of the α-phase and β-phase reflections, zone axis <331> with diffuse strands; *D* is the β-phase grain (dark field); *E* - microdiffraction of β-grain, zone axis <130>.

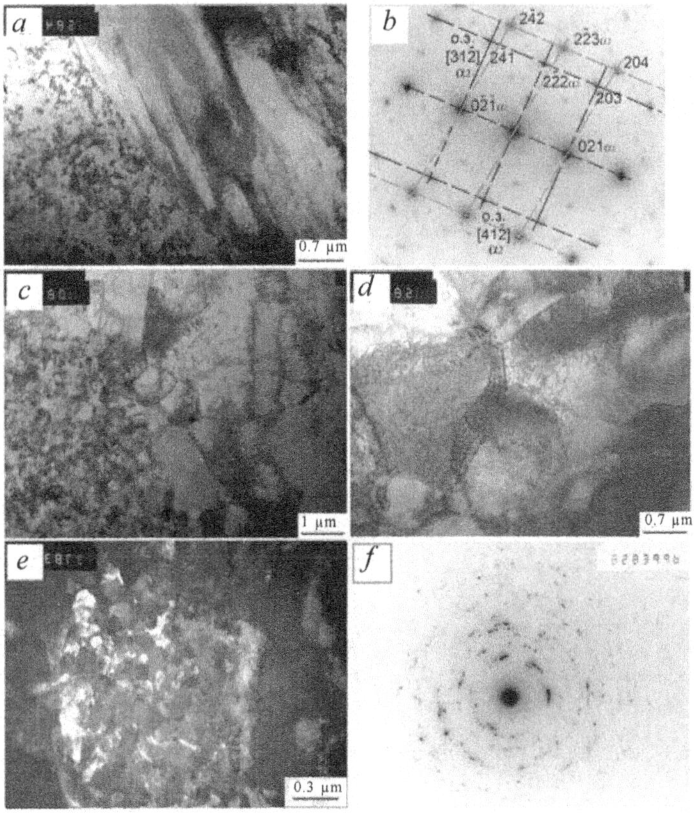

Fig. 17. Welded joints ($\mathbf{A}_w$), the microstructure of the transition zone and adjacent areas: *a* – banded structure of aluminide (dark field); *b* – microdiffraction of aluminide (Fig. 10 *a*), cross sections of inverse lattices $[31\bar{2}]\alpha_2$ and $[41\bar{2}]\alpha_2$; *c* – recrystallized zone of titanium; *d* – dislocations inside titanium grains; *e* – nanostructure of the vortex zone, *f* – microdiffraction of the nanostructure.

On one side of the interface there is the titanium aluminide, having a strip structure (Fig. 17 *a, b*), and on the other side titanium (Fig. 17 *c*). Microdiffraction was obtained for the region of the strip structure and contains superstructural reflections of the aluminide (Fig. 17 *b*).

In Fig. 17 *c*, the transition zone and the recrystallization zone of titanium are visible at the same time. The titanium grains contain a high dislocation density, preserved during recrystallization (Fig. 17 *d*). The transition zone consists of grains that are much smaller (more than an order of magnitude) than the grains that form the recrystallization zone of titanium. In the micrographs shown in Fig.

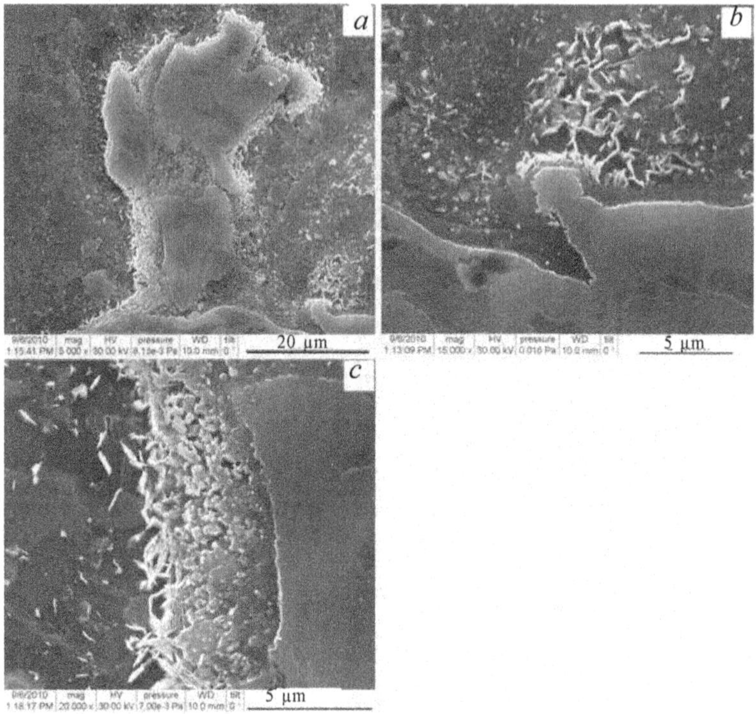

Fig. 18. Scattering of aluminide particles for the welded joint ($\mathbf{A}_w$): *a...c* – different magnifications.

16, 17, the grains in the transition zone have sizes 50...300 nm, with the exception of some large grains. The selected area in Fig. 17 *d* contains α-grains with sizes of about 30...50 nm, i.e. in fact is nanocrystalline.

Thus, the local melting is confirmed by the observation of the duplex structure (Figs. 16, 17), consisting of clearly-cut grains of the α- and β-phases of variable composition, and the β-phase was not observed at all in the original materials in the initial state.

The expansion of micron particles of aluminide, which is clearly visible in Fig. 18, is surprisingly similar to the flying out of fragments of a different size, which occurs in an explosion. In Fig. 18 it can be seen that particles of arbitrary shape fly out of the area occupied by the aluminide. But it remains a fragmented layer adjacent to the border with titanium. This layer is especially clearly seen in the micrograph (Fig. 18 *c*), obtained with at a high magnification.

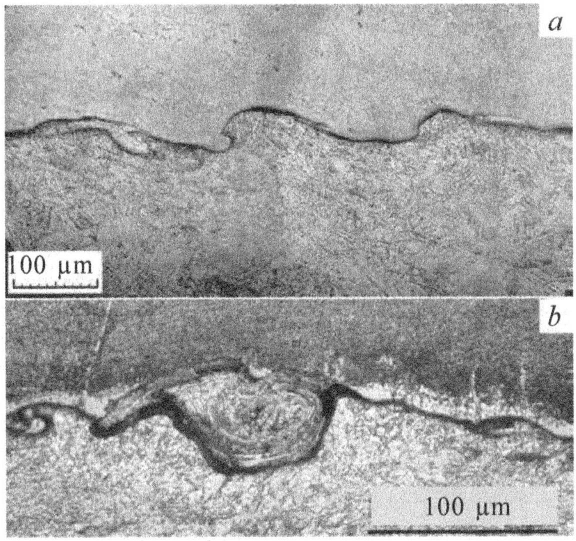

Fig. 19. Cross-section of the wave-shaped interface for the welded joint ($\mathbf{B}_w$): *a*, *b* – with different magnification (optical micrograph).

The micrographs shown in Fig. 18 play a special role since they are in fact evidence of the fragmentation process of the type of fragmentation proposed by us and discussed further in chapter 4.

3.1.2. B_w welded joint: titanium VTI-4, the wavy interface

The welded joint $\mathbf{B}_w$ was produced using VTI-4 alloy, having the composition: Ti–21.9 Al–23.5 Nb–0.9 Zr–0.45 Mo–1.39 V (at.%).

Figure 19 shows the optical image of the wavy interface. The wavelength and amplitude are ~250 μm and ~30...60 μm, respectively. Let us pay attention to the different shape of the wave-like boundaries shown in Fig. 7 and 19 *a*. The boundary in Fig. 19 *a* is the so-called asymmetric boundary [1]. Figure 19 shows the isolated regions of the melt are visible, as in Fig. 7. In Fig. 19 *b* the vortex-shaped structure of the zone is clearly visible. Figure 20 shows for both welded joints shows the dispersion of aluminide particles near the interface.

Data on the chemical composition of the vortex zone for the welded joint $\mathbf{B}_w$ were obtained using SEM by numerous measurements, just as above for the $\mathbf{A}_w$ welded joint. As can be seen from Table 3, the chemical composition of the zone differs from the

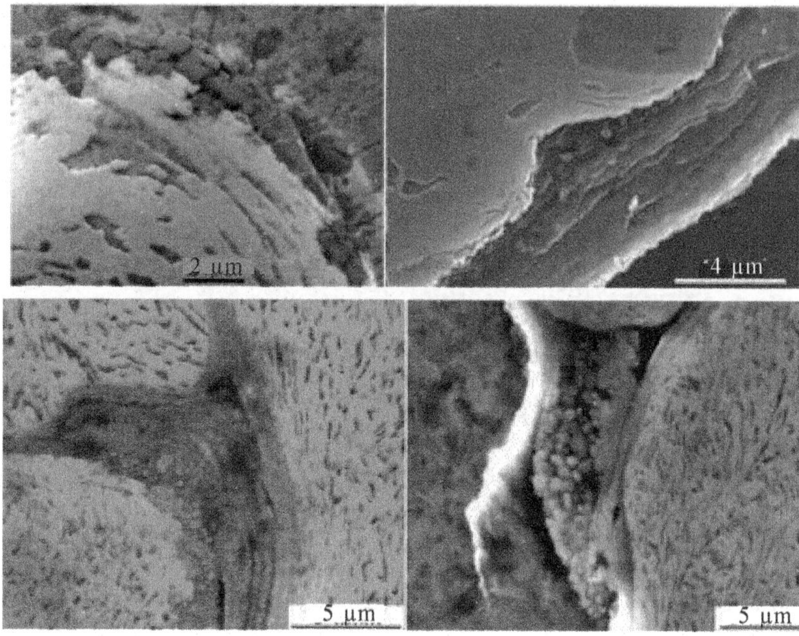

Fig. 20 a. Dispersion of aluminide particles near the VTI-1–titanium interface (SEM): particle penetration into the melt zone
Fig. 20 b (right). Dispersion of aluminide particles near the VTI-4–titanium interface (SEM).

Table 3. Results of spot measurements of chemical compositions of the vortex zone (VTI-4–titanium) at.%

Measurement No.	Concentration, at.%		
	Al	Nb	Ti
1	0.71	0.61	98.68
2	1.6	1.05	97.35
3	8.30	9.15	82.55
4	6.91	8.77	84.32
5			100.00

composition of the starting metals: aluminide (Ti–21.9 Al–23.5 Nb for basic elements) and titanium.

In fact, the main part of the vortex zone contains a solid solution based on titanium. Such a composition indicates the penetration of elements from one material into another, which was observed above for the vortices in the transition zone of the welded joint (VTI-1–titanium).

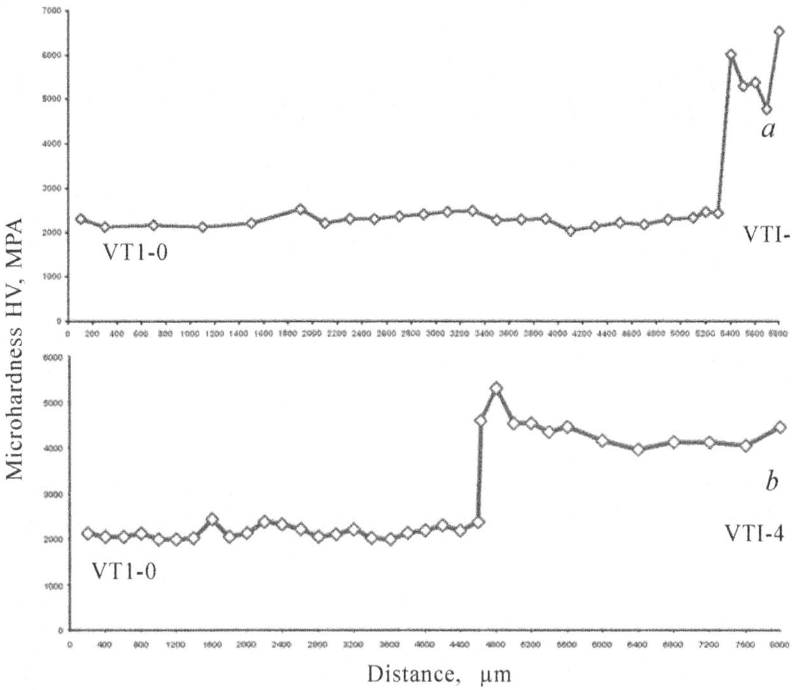

Fig. 21. Microhardness depending on the distance from the interface: *a* – VTI-1-titanium, *b* – VTI-4–titanium.

Certification tests were carried out for the welded joint VTI-4– titanium at the CRISM 'Prometey' (St. Petersburg). The results of the test reports for shear and bending are shown below (Table 4). It was found that the fracture always occurs in titanium. The shear strength was higher than 350 MPa. Depending on the type (external or internal) of the aluminide layer, the bending angle is approximately either 10° or 25°.

In Table 4 b_0, h_0, F_0 are the sides and cross-sectional area of the cladding layer, P_{max} is the maximum load, σ_{shear} is the ultimate shear strength.

Table 4. Shear test results

Sample	b_0, mm	h_0, mm	F_0, mm	P_{max}, N	σ_{shear}, MPa
1	9.48	5.28	50.05	18700	374.5
2	9.75	5.29	51.58	19000	367.6

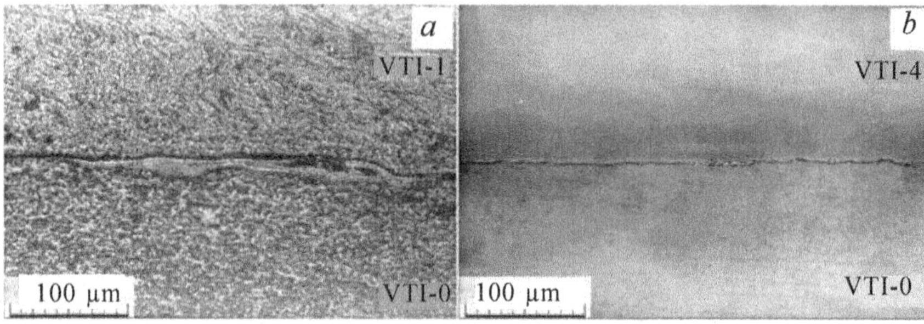

Fig. 22. Interfaces (cross section), optical micrograph: *a* – welded joint $\mathbf{A}_p$); *b* – welded joint $\mathbf{B}_p$.

In addition, microhardness measurements were performed for the welded joints ($\mathbf{A}_w$) and ($\mathbf{B}_w$) , the results are presented in Fig. 21.

As can be seen from Fig. 21, the microhardness in the transition zone has high values: about 6500 MPa for the $\mathbf{A}_w$ welded joint and 5500 MPa for the $\mathbf{B}_w$ welded joint.

3.1.3. (A_p) welded joint: titanium–VTI-1, flat melted interface

Figure 22 shows an optical image of the interfaces for the $\mathbf{A}_p$ and $\mathbf{B}_p$ welded joints. The interface in Fig. 22 *a* does not contain waves, and that in Fig. 22 *b* contains irregular waves with a very small amplitude.

Below it will be shown that along these boundaries there is complete or partial melting, respectively. For convenience, hereafter we call them melted or partially melted boundaries, although, of course, solidification occurred.

Figure 23 shows the internal structure of the fully melted (and then solidified) interface for the $\mathbf{A}_p$ welded joint at different

Table 5. Results of spot measurements of the chemical composition near the interface for the $\mathbf{A}_p$ welded joint

Measurement No.	Concentration, at.%		
	Al	Nb	Ti
1	2.40	1.72	95.89
2	16.68	11.78	71.54
3	22.22	14.23	63.55
4	22.39	15.25	62.35
5	21.02	13.27	65.70.

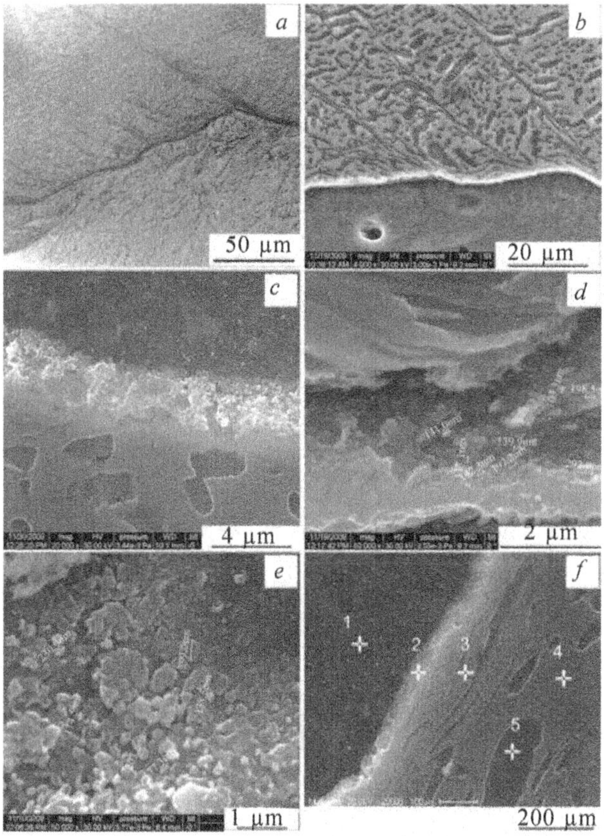

Fig. 23. Structure (SEM) of the interface for the $\mathbf{A}_p$ welded joint (VTI-1–titanium): *a, b* – shear bands in the aluminide; *c–e* – the internal structure of the interface, *f* – the measurement points of the welded joint near the interface.

magnifications. The alu))minide in Fig. 23 *a, b* shows clearly visible shear bands (plastic deformation localization bands). As can be seen from Fig. 23 *a, b*, the shear bands are located at an angle to the interface. This contradicts the assumption [31] that the plastic deformation localization bands are located along the interface. In Fig. 23 *b*, the border is visible as a melted 'bundle', the thickness of which varies within 10...30 μm. As can be seen from Fig. 23 *d, e*, the structure of the fully molten boundary is ultrafine: the dimensions of the spherulites are 100...400 nm.

Data on the chemical composition obtained using SEM in numerous measurements, which were made for the interface are shown in Fig. 23 *f*. In Fig. 23 *f* to the right of the interface there is

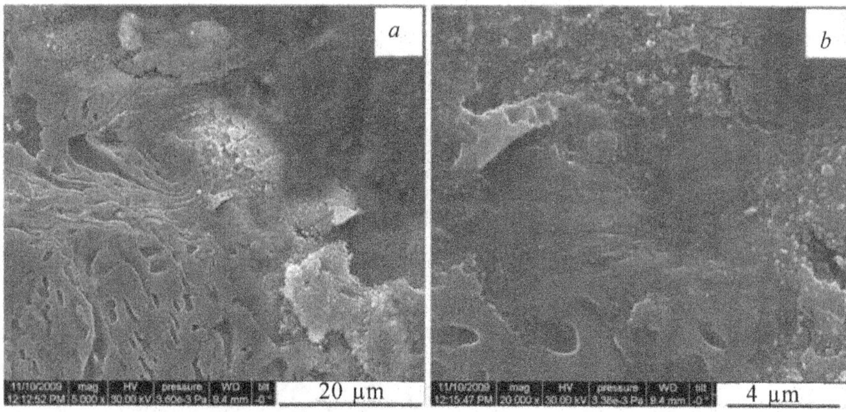

Fig. 24. Breakthrough of plastic flow across the interface for the $\mathbf{A}_p$ welded joint VTI-1–titanium (SEM): *a* – transition through the interface (on the left – aluminide, on the right – titanium); *b* - fragment of Fig. 24 *a* at higher magnification.

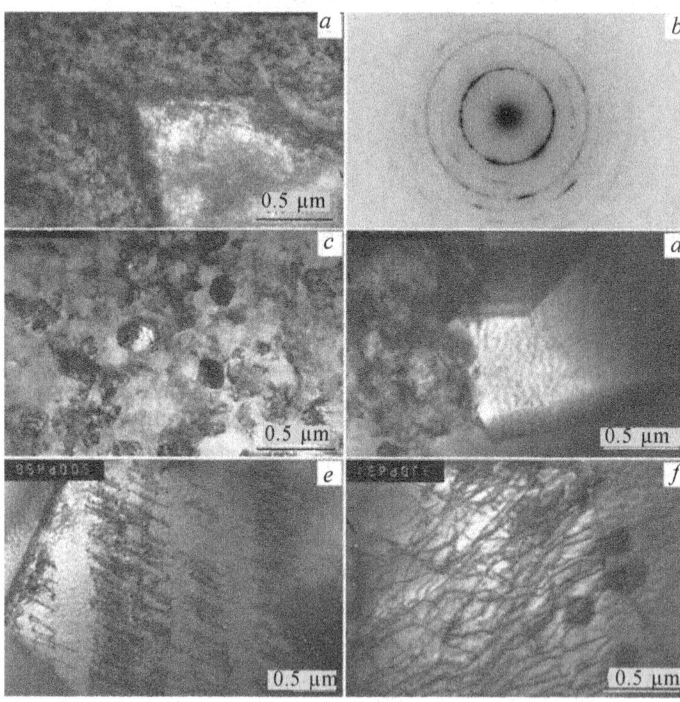

Fig. 25. TEM images of the interface for the $\mathbf{A}_p$ welded joint (VTI-1–titanium): *a* – ultrafine β-phase structure; the bright region is the orthorhombic aluminide; *b* – microdiffraction from the area shown in Fig. 25 *a*; *c*, *d* – larger β-phase grains; *e*, *f* – dislocations in β-phase grains.

the aluminide (globules are visible), to the left is titanium. Table 5 shows the concentrations of elements.

As can be seen from Tab. 5, the chemical composition differs from the composition of the starting metals: aluminide (Ti - 30 Al - 16 Nb for basic elements) and titanium. Such a composition indicates the penetration of elements from one material into another. In fact, the main part of the transition zone contains a solid solution, the composition of which differs from the middle one and is closer to the composition of the aluminide.

The micrograph in Fig. 24 *a* reflects the dynamics of the process of breaking of the plastic flow across the interface, as a result of which a part of the interface was inside the titanium band. We draw attention to the ultrafine structure of the interface, which is clearly visible in Fig. 24 *b*. As shown by TEM analysis (Fig. 25), the interface under study consists mainly of the disordered β-phase (BCC). In some areas, grains with a size of 10...30 nm are formed (Fig. 25 *a*). Electron diffraction patterns from such areas have the form of rings, which consist of individual reflections of high density (Fig. 25 *b*).

The calculation showed that the structure shown in Fig. 25 *a* indeed corresponds to the β-phase. Nevertheless, in some electron diffraction patterns very weak β-phase reflections (HCP) are recorded. Due to the fact that the width of the molten layer is not constant along the entire surface of the joint, the heat sink does not occur uniformly during cooling. The structure of the formed phase is heterogeneous. There are areas with larger grains and clear boundaries. The grain size is approximately 50...100 nm (see Fig. 25 *c*). Electron diffraction analysis showed that these grains are also β-phases. Figure 25 *d* shows grains of the β-phase the size of which exceeds 500 nm. In Fig. 25 *e*, *f* both rectilinear and curvilinear dislocations are visible.

It is believed that the observation of dislocations in recrystallized grains [32] becomes possible due to the high cooling rate: grains form from many centres arise and then upon subsequent cooling the growth of individual grains that absorb some grains and exert pressure on others leads to the formation of dislocations due to constrained deformation.

3.1.4. (B_p) welded joint titanium–VTI-4, almost flat, partially melted interface

Let us proceed to the analysis of the transition zone of the $\mathbf{B}_p$ welded joint (VTI-4–titanium), the cross section of which is shown in Fig. 22 *b*. The width of the interface (approximately 5 μm) is much smaller than in the case of the completely melted interface considered above. For the thin interface investigated here, although it is corrugated, it

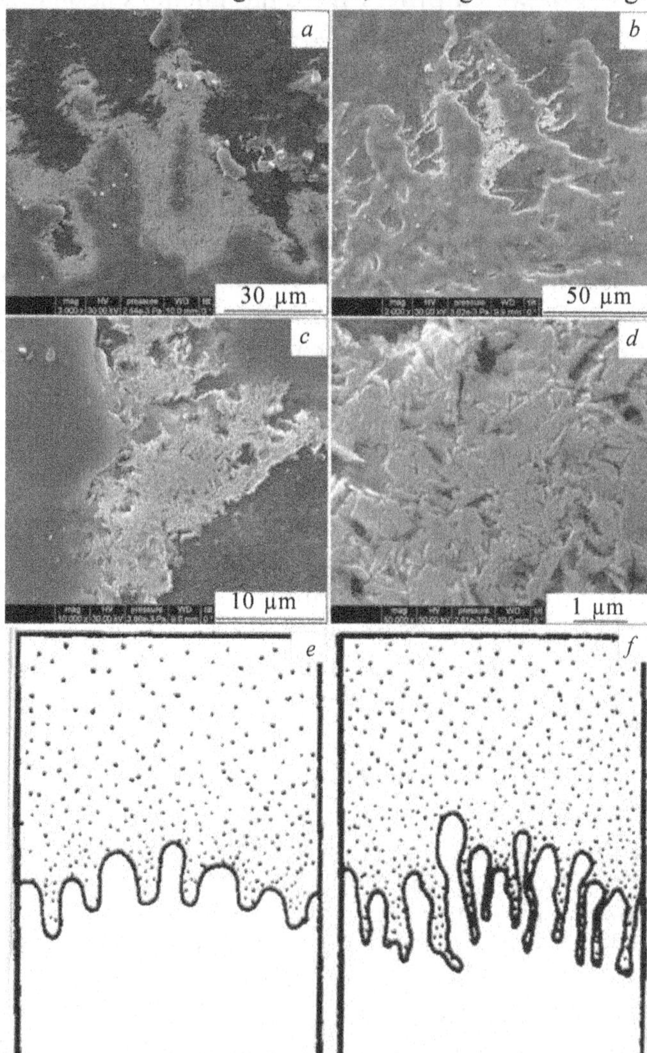

Fig. 26. The structure of the partially molten interface ($\mathbf{B}_p$) (VTI-4–titanium): *a*, *b* – interface at low magnification; *c...d* – dendritic structure, *e*, *f* – 'sticky fingers'.

is not possible to use longitudinal sections, because due to the small width of the interface, the longitudinal section in one case turns out to be in the aluminide, and in the other in titanium. Therefore, we use inclined sections. Figures 26 *a, b* show the interface at low magnification.

Instead of the alternation of bands observed in the longitudinal section of the corrugated surface (see, for example, Fig. 8), here is a different picture: the bottom of Fig. 26 *a, b* contains an aluminide band, the top – a band of titanium, with alternating 'fingers' of both materials in the transition zone. The 'frosting' interface on the titanium side is clearly visible. Based on Figs. 26 *b...d* it is supposed that this is a dendritic structure. At a high magnification (Fig. 26 *d*), it is clear that the structure is formed by 'fingers', but an order of magnitude thinner than in Figs. 26 *a, b*. Perhaps this is a manifestation of the self-similarity of the structure at various stages of its development, so that the micrographs in Fig. 26 are images of fractal structures. Their similarity with the images of fractal structures shown in [33], where the dendritic structure is the same as in Fig. 26, is not related to certain crystallographic directions.

Figures 26 *e, f* show the image of 'sticky fingers' [33]. They result from the interpenetration of liquids with different viscosities. In this case, it is an oil and an aqueous solution of glycerin. It is unclear whether the structure observed in Figs. 26 *b...g* are structures with 'sticky fingers', although the aluminide does indeed have a greater viscosity than titanium, because it contains α_2-phase globules.

Thus, various structures indicate a meltdown for both studied interfaces. For a completely molten interface (Fig. 23), this structure contains spherulites. For a partially molten interface (Fig. 26), this is a dendritic structure. It was already mentioned above that the orthorhombic alloy, although considered to be an intermetallic compound, at certain compositions and temperatures, can turn into a disordered-phase (BCC). It is possible that such a transformation is realized when the melt solidifies. The above mentioned properties of the β-phase, such as increased plasticity and viscosity, provide good quality of the welded joint under investigation, despite the fact that when it is formed, melting occurs along the interface, full or partial.

As a result, the explosion-minded stereotype, i.e. the danger of melting, is overcome. Accordingly, it is impossible to consider explosive welding to be cold welding. Moreover, it is possible that the molten layer promotes adhesion, facilitating the interpenetration of materials. Obviously, the molten layer should be quite thin. In addition, there should be no 'bad' phases. A detailed analysis of possible meltdown scenarios and associated potential risk zones will be carried out in Chapters 4 and 5.

3.2. Copper–tantalum

Mixing is a complex task, in particular for metals that do not have mutual solubility. In order to find out how important the presence of the mutual solubility of the starting materials is, metals (copper–tantalum), which under normal conditions do not have mutual solubility, are selected for explosive welding, and form immiscible suspensions in the liquid state.

The copper–tantalum welded joint is the most thoroughly studied by the authors of the monograph. It can be said say it is also the most 'favorite' for some of them. The results are published in [34–36], as well as the reviews cited above [10–14]. In this case, the inhomogeneities of the interface for the copper–tantalum welded joint are discussed in [11], fragmentation in [12], dissipative processes in [13]. It is the copper–tantalum welded joint (**C**$_w$), which has a wave-like interface, is part of the chemical reactor wall [37], the reasons for the high quality and stability of which will be discussed further in section 5.1.

3.2.1. (C_w): copper-tantalum welded joint, flat interface

Tantalum (TVCh) and copper (M1) were chosen as the starting materials. Welding was performed in the Khimmash company, using parallel arrangement of the plates. The tantalum plate thickness was 1 mm, the copper plate thickness was 4 mm, the gap between the plates was 1 mm. The copper plate was thrown on tantalum, which lay on a titanium and steel backing plate with a thickness of 4.5 + 20 mm. The detonation velocity V_d = 2680 m/s. The impact of the plates occurred at an angle $\gamma = 5.22°$ with velocity of V_c = 234 m/s. The choice of the parameters corresponds to the lower limit of weldability. Such a regime, traditionally used by the Khimmash company, is the most economical because of the lower charge and, accordingly, lower costs for explosives. It is also significant that in this case the impact of the shock wave on the surrounding objects decreases.

Figure 27 shows the cross section of the transition zone for the **C**$_p$ welded joint. It is clearly seen that the interface is almost flat

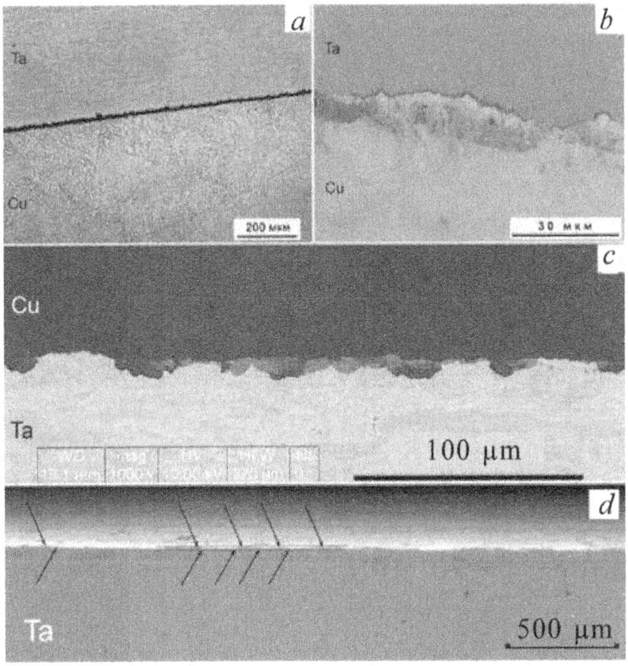

Fig. 27. Cross section of the Cu–Ta interface (**C**p): *a, b* – optical micrograph, with different magnification; *c* – SEM, *d* – surface of the original tantalum plate.

Table 6. Results of spot measurements of chemical composition near the interface for the $\mathbf{C}_p$ welded joint

Measurement No.	Concentration, at.%	
	Cu	Ta
1	2	98
2	74	26
3	83	17
4	97	4
5	100	0
6	1	99
7	86	14

(Fig. 27 *a*). As can be seen from Fig. 27 *b, c* for the Cu–Ta welded joint, the interface, as for the titanium–aluminide welded joint studied above, is not smooth, but contains cusps. The dimensions of the cusps (Fig. 27 *c*) are approximately 5...10 μm. Earlier, the cusps were found in the transition zone of the titanium–orthorhombic titanium aluminide welded joint (Fig. 10).

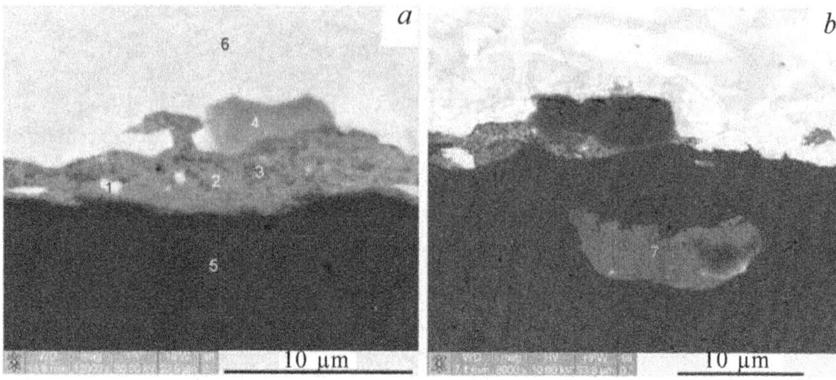

Fig. 28. SEM image of the cross section of the (**C**p) Cu–Ta interface in backscattered electrons (EBSD): 1, 6 – tantalum, 2, 3, 7 – a mixture of copper and tantalum, 4, 5 – copper.

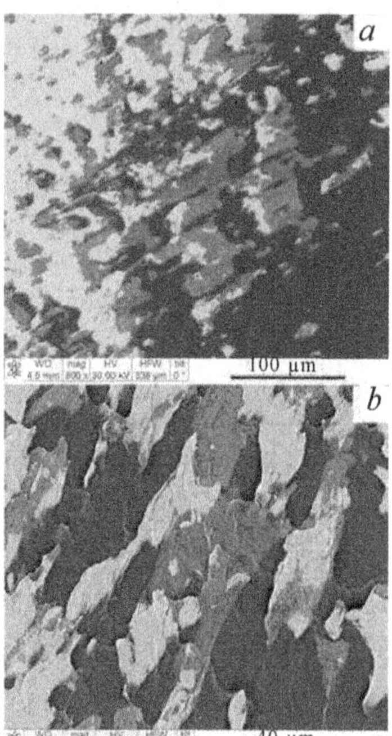

Fig. 29. Longitudinal section of the flat interface (Cp) Cu - Ta with different magnification (SEM): white spots - Ta; black spots - Cu; gray spots - a mixture of Ta and Cu

For the copper–tantalum C_p welded joint, the surface roughness of metals prepared for welding was measured using a profilometer. The average roughness is about 0.610 µm for tantalum and 0.410 µm for copper. This means that the size of the cusps after welding is 10–20 times greater than the initial roughness. Let us emphasize

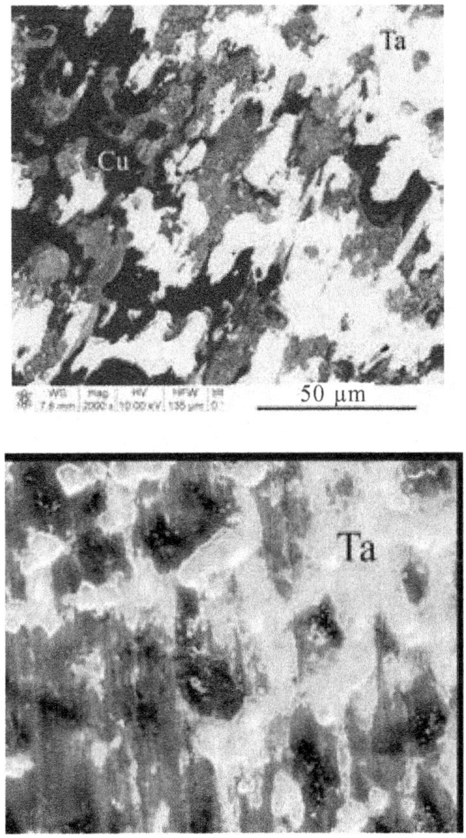

Fig. 30. Longitudinal section of the transition zone: *a* – welded joint ($\mathbf{C}_p$), areas of tantalum, copper and local melting zone; *b* – welded joint ($\mathbf{E}_p$), areas of tantalum, aluminium and local melting zone.

the essential point: on some tantalum cusps, as can be seen from Fig. 27 *c*, smaller cusps of the following orders are formed.

The SEM image of the cross section (Fig. 28) clearly shows the alternation of areas of the transition zone with different concentrations. The results of concentration measurements at the points indicated in Fig. 28 are shown in Table 6. Areas of three colors are observed, corresponding to both metals (tantalum and copper) and their mixtures.

The same areas are visible on the SEM image of a longitudinal section (Fig. 29). Instead of the image which for a wavy boundary would consist of parallel bands (see, for example, Fig. 8), we see an image consisting of spots of three colours: white, black and gray.

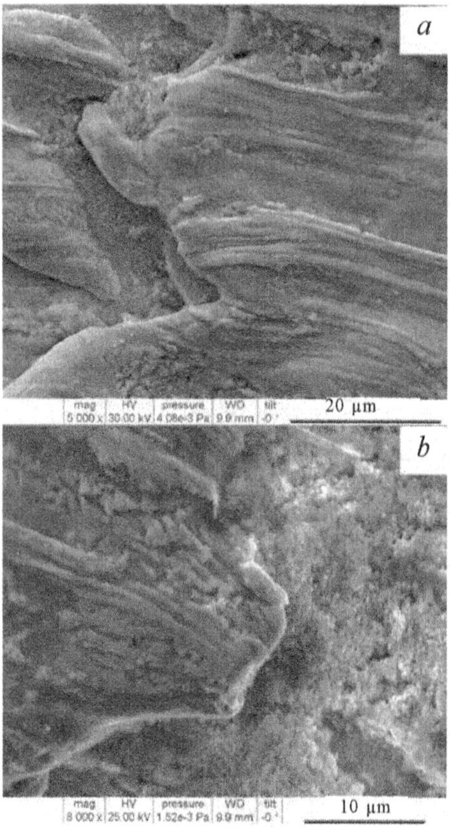

Fig. 31. Welded joint (**C_p**), longitudinal section (SEM): copper etched: *a, b* – with different magnification.

This means that the transition zone consists of randomly distributed regions of three types: the tantalum zone corresponds to white, the copper zone to black, and the local melting zone to gray. In Fig. 29 we can see the stretching of spots along a certain chosen direction.

The reason for the observation of the tricolour nature of the longitudinal section are the cusps on the interface. If there were no cusps, then one metal would be observed on the longitudinal sections above the interface, and another one below, so that the SEM images would be monochromatic. In the presence of cusps, as a result of their intersection by a plane of longitudinal section, regions containing different metals would be observed simultaneously, so that the image would be two-colour. However, in Fig. 29 there are not two, but three areas. In addition to the areas filled with the initial metals, there is a gray local melting zone containing a mixture of both metals.

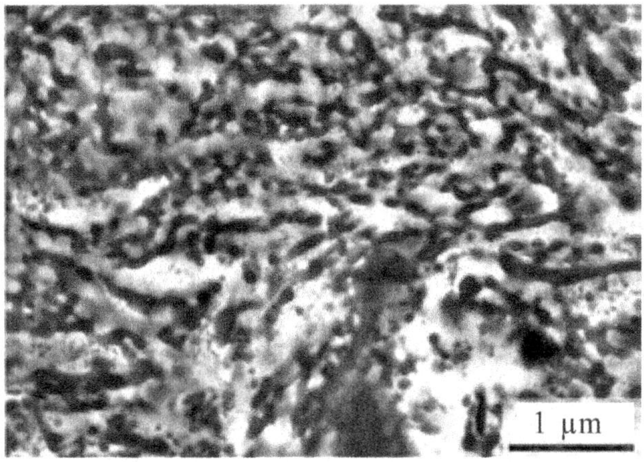

Fig. 32. Microheterogeneous structure of the local melting zone (SEM) for the $\mathbf{C}_p$ welded joint.

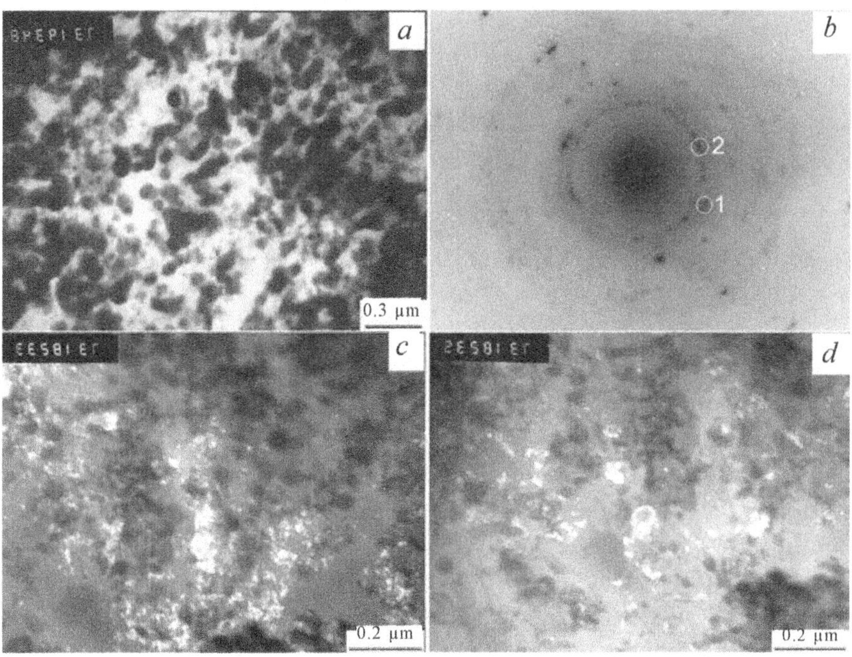

Fig. 33. The microstructure of the gray zone (TEM) for the $\mathbf{C}_p$ welded joint: *a* – bright-field image; *b* – microdiffraction; *c* – dark-field image (for Fig. 33 *a*) in the <111>Cu reflex, *d* – dark-field image (for Fig. 33 *a*) in the <110>Ta reflex.

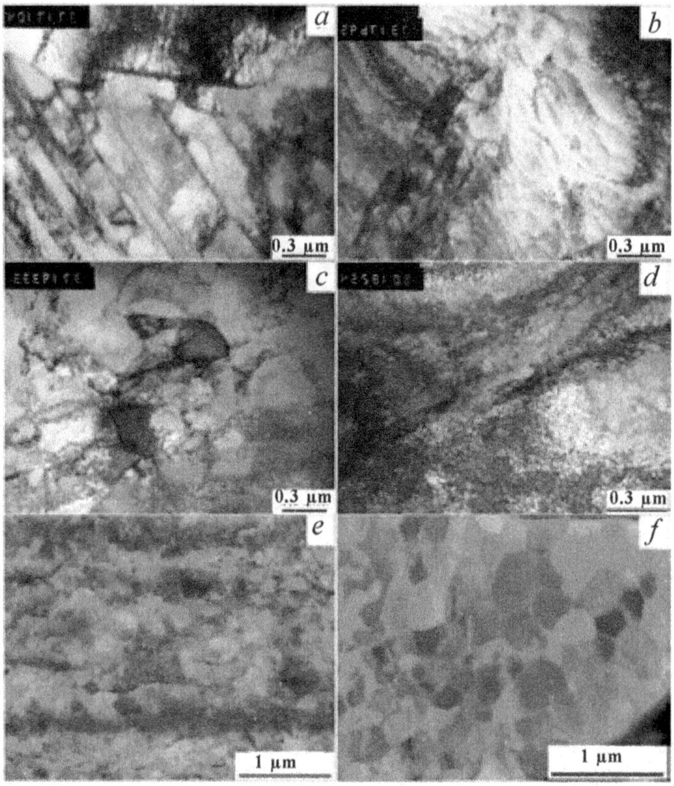

Fig. 34. C_p welded joint. TEM image of the deformed structure of copper and tantalum: band structure of copper (*a*) and tantalum (*b*); copper cellular structure (*c*); strong cold working in tantalum (*d*); recrystallized regions in copper (*e*) and tantalum (*f*).

As a result, the interface is a chaotic relief with a large number of cusps and depressions. For clarity there is Fig. 30, where tricolour images are not only convincing, but also impressive. The results of studying the $\mathbf{E}_p$ aluminium–tantalum welded joint are presented in Section 3.3.

To clarify the structure of the gray zone (Fig. 29), Fig. 31 shows a longitudinal section of the transition zone, obtained after the copper has been completely etched. On the surface of tantalum there are visible tantalum particles, which are mainly of the nanometric size.

The TEM study of the Cu–Ta welded joint is difficult, since for such dissimilar materials the selection of reagents in the manufacture of foils is rather difficult. Under the action of reagents suitable for tantalum, copper can be completely etched. In this paper, foils were prepared using an ion cannon.

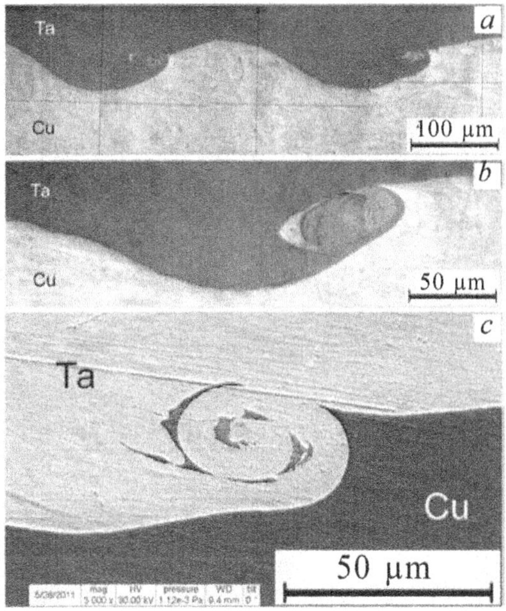

Fig. 35. Cu–Ta welded joint ($\mathbf{C}_w$), the cross section of the wavy interface: *a* is the interface, *b* is the zone of local melting near the boundary (optical micrograph); *c* is a macroturn in the tantalum zone (SEM).

The microheterogeneous structure of the local melting zone is clearly visible both on the SEM (Fig. 32) and on the TEM images (Fig. 33). In Fig. 33, and a bright field image of the microstructure of the local melting zone is shown. In Fig. 33 *c*, the dark field image is shown in the <111> Cu reflex, and in Fig. 33 *d* – the dark-field image in the <110>Ta reflex. In Fig. 33 the Ta particles are dark. The electron diffraction pattern (Fig. 33 *b*) shows a system of rings consisting of individual reflexes the decoding of which showed that they belong to tantalum. Strong spot reflexes are reflections from copper. The type of electron diffraction pattern indicates a significant disorientation of both Ta and Cu particles. It can be assumed that exactly the tantalum nanoparticles observed here remain on the tantalum surface after copper is etched (Fig. 31).

In addition to the gray zone, the transition zone contains the copper and tantalum zones mentioned above (Fig. 29). As shown by TEM analysis, these zones do not undergo melting and have a typical structure for severe plastic deformation. Figure 34 shows images of the highly deformed copper structure (left) and tantalum (right) for the $\mathbf{C}_p$ welded joint. In both materials, there is a band structure (Fig.

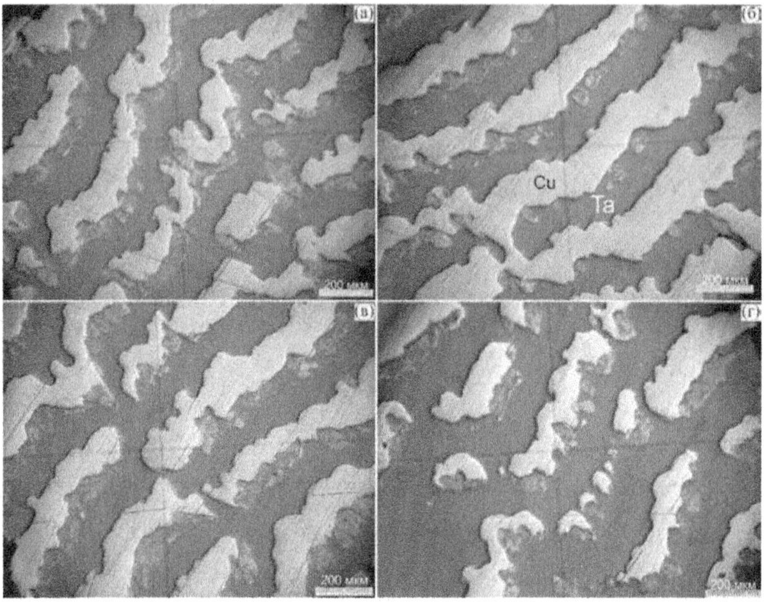

Fig. 36. Longitudinal section of the wave-shaped interface for the Cu–Ta welded joint (C_w): panorama (optical micrograph).

34 *a, b*), a cellular structure (Fig. 34 *c*), a high dislocation density (Fig. 34 *d*) and recrystallized regions (Fig. 34 *d, e*). The grain size in these areas is approximately 100...300 nm, which is several orders of magnitude smaller than the original size (approximately 100 μm).

Similar structures were observed above for the Aw titanium–aluminide welded joint (Fig. 12). These are typical structures for fragmentation under severe plastic deformation (see Section 4.2).

3.2.2. (Cw): copper–tantalum, wavy boundary

Laboratory samples were used for welding: tantalum (thickness 0.1 mm) and copper (thickness 3.5 mm), the remaining dimensions are tens of mm. As in the preparation of the above-considered C_p compound, the same scheme was used here: a parallel arrangement of the plates, a fixed plate on a metal substrate. Welding was carried out at the Volgograd State Technical University. We confine ourselves to giving here only the basic welding parameters for obtaining a (C_w) welded joint: $\gamma = 11.8°$, $V_{con} = 2125$ m/s, $V_c = 440$ m/s (V_{con} – the

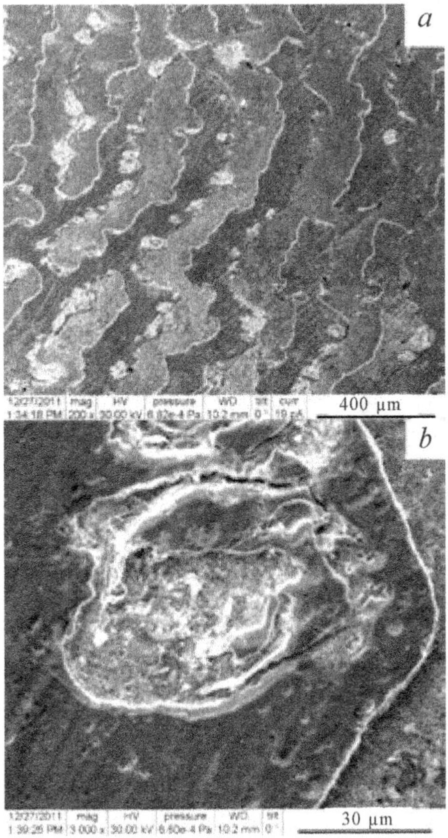

Fig. 37. Longitudinal section of the wavy interface for the welded joint ($\mathbf{C}_w$) Cu–Ta (SEM): *a* – copper and tantalum bands, local melting zones, cusps; *b* – vortex structure of the local melting zone.

velocity of the point of contact, V_c – the collision velocity)

As can be seen from Fig. 35 *a*, the interface in this case is wavy and in shape close to the one shown in Fig. 19, and for the welded joint ($\mathbf{B}_w$). In both cases, the wave-like profile of the interface has asymmetry about the vertical plane. The wavelength and amplitude in Fig. 35 equal approximately ~270–350 μm and ~60–65 μm respectively. In Fig. 35 *b* the zone of local melting is clearly visible. In Fig. 35 macroturns are visible in the tantalum zone. In contrast to the vortices in the liquid phase, the macroturns are formed in the solid phase due to severe plastic deformation. Similar macroturns were observed for the $\mathbf{A}_w$ welded joint in titanium aluminide (Figs. 9, 10, 11 *c*).

A longitudinal section of the interface is shown as a panorama in Fig. 36 (optical micrograph). There are projections, the size of which

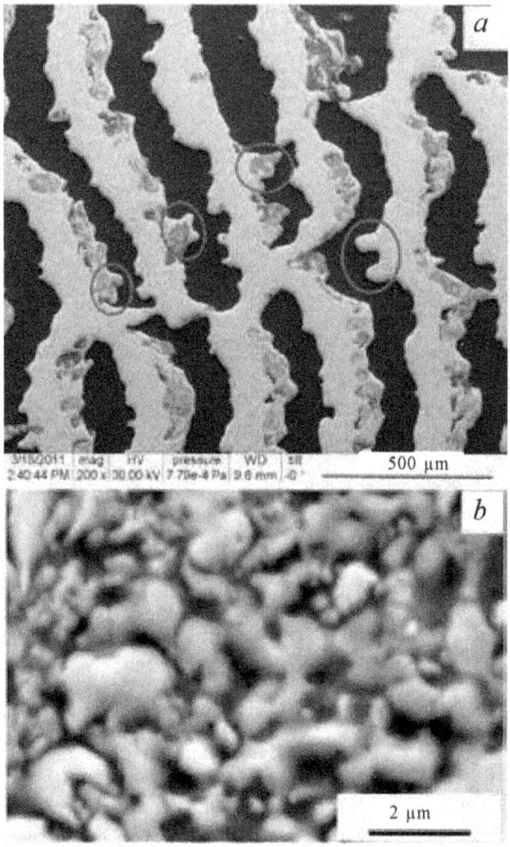

Fig. 38. Longitudinal section of the interface for the ($\mathbf{C}_w$) Cu–Ta welded joint (SEM): *a* – cusps and cusps on the cusps; *b* – the internal structure of the local melting zone.

is several tens of microns. The system of strips is not regular as a result of the breakage of some strips and branching of others. The parts that appear during the decay of the strips are visible: parts of one strip are surrounded on all sides by another strip. The closure of the bands can be seen.

The SEM image of the longitudinal section is shown in Fig. 37 *a*. It is obtained without the use of an ion cannon and the image quality is not high enough. However, in Fig. 37 *a* numerous zones of local melting (white colour) are clearly visible, and in Fig. 37 *b* – the vortex structure of one of the zones.

In contrast to the vortices considered above, observed for the titanium–aluminide welded joint ($\mathbf{A}_w$), in this case (Fig. 37 *b*) it is an imperfect vortex structure.

In Fig. 38 *a*, the inhomogeneities of the interface mentioned above are visible: cusps and local melting zones. In addition, some of the

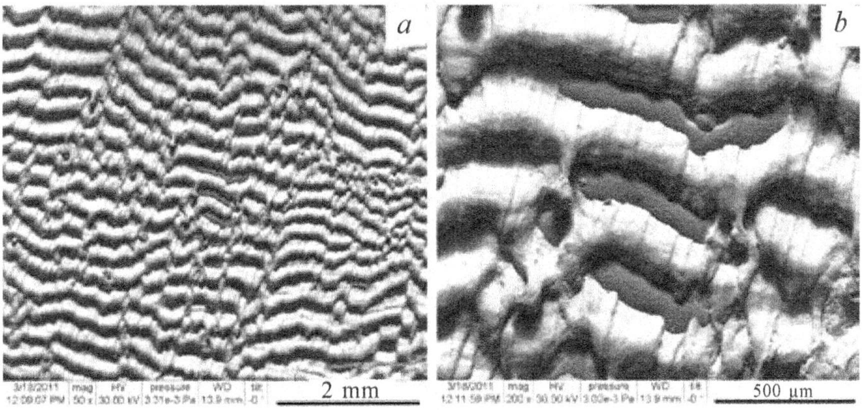

Fig. 39. Welded joint (C_w), SEM image of the longitudinal section (copper was etched out): *a*, *b* – at different magnifications.

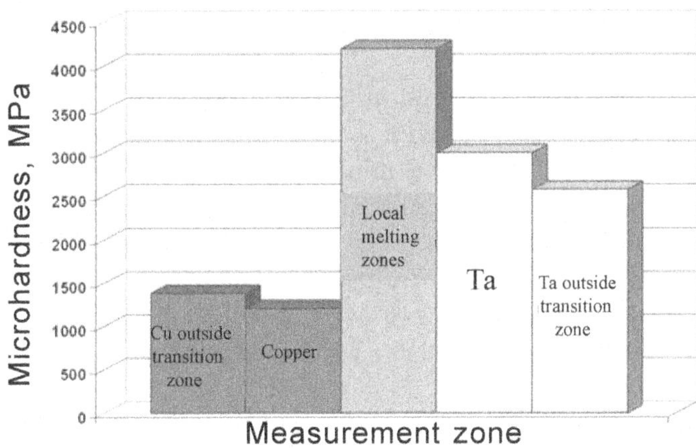

Fig. 40. Microhardness in different areas inside and outside the transition zone of the welded joint (C_p).

cusps show the formation of smaller cusps of the following orders (circled in red in Fig. 38 *a*). One can clearly see how imperfect the interface is. The observed decay and closure of the bands is due to the formation of large cusps.

The SEM image of the internal structure of the local melting zone is shown in Fig. 38 *b*. As can be seen from the comparison of Fig. 38 *b* and Fig. 33, for both copper–tantalum welded joints, regardless of the interface shape, the local melting zone is filled with a copper matrix (solidified after melting) and tantalum particles not experiencing melting. In the case of a flat interface, these are tantalum nanoparticles, in the case of a wave-like micron particles.

Some of these features of the interface are clearly visible (Fig. 39) on a more prominent SEM image of a longitudinal section obtained after the copper has been completely etched. Here, instead of bands, the crests of tantalum are visible. In Fig. 39, and it is clear that the direction of the crests changes when moving from one area to another. In Fig. 39 *b* the link between the tantalum crests is visible, leading to the formation of complex three-dimensional defects, which substantially distort the interface surface. It is clearly visible how imperfect it is.

Figure 40 shows the microhardness values both in the transition zone and outside the zone. It can be seen that the microhardness values of tantalum and copper in the transition zone changed only slightly compared with the corresponding values outside the transition zone. But in the mixing zone the microhardness is much higher. The difference in the particle sizes of the dispersed phase, i.e. tantalum leads to different microhardness values measured in localized melting zones.

For the compound (Cp), the microhardness in this zone is about 4000 MPa, which is 1000 MPa higher than the microhardness of tantalum and 3000 MPa - the microhardness of copper. However, for the compound (Cw), the microhardness in this zone practically coincides with the microhardness of tantalum. As shown by mechanical tests, the shear strength is also higher for the welded joint (**C**$_p$) and is about 230...240 MPa, whereas for the welded joint (**C**$_w$) it is about 150...170 MPa, which, nevertheless, exceeds the strength of copper.

Extremely useful was the use of different welding modes for the same copper–tantalum pair. The section surface has a different shape: flat in one case, wavy in the other. But in any case, the interface is not smooth: it contains cusps of tantalum and local melting zones.

3.3. Aluminium–tantalum

The welded joints (**E**$_p$) and (**E**$_w$) of aluminium–tantalum metals having mutual solubility were studied. As can be seen from Fig. 1, the parameters for the welded joints (**E**$_p$) and (**C**$_p$) having flat interfaces are located near the lower weldability limit. If we compare the welded joints (**E**$_w$) and (**C**$_w$), then the welding parameters for the

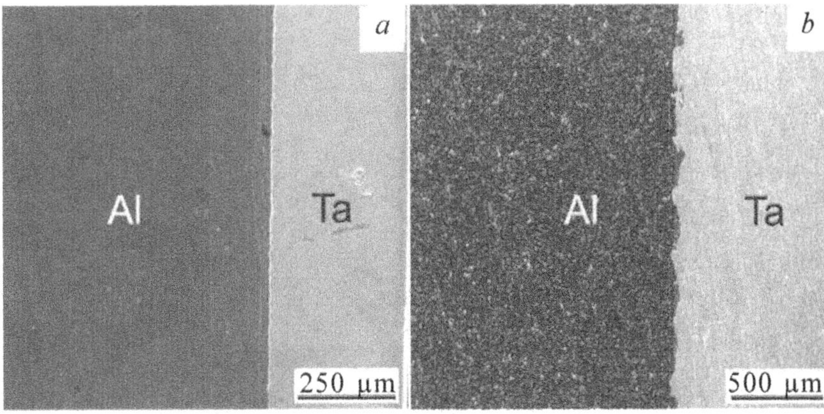

Fig. 41. The interface (cross section) for aluminium–tantalum (SEM) welded joints: *a* – welded joint ($\mathbf{E}_p$); *b* – welded joint ($\mathbf{E}_w$).

welded joint ($\mathbf{E}_w$) are significantly higher.

For the welded joint ($\mathbf{E}_p$): $\gamma = 8.6°$, $V_{con} = 2000$ m/s, $V_c = 300$ m/s. The arrangement of the plates is parallel, the aluminium plate was thrown on the tantalum plate, the thickness of the tantalum plate was 0.5 mm, the aluminium plate 7 mm. The tantalum plate fits snugly to a steel plate with a thickness of 15.5 mm. When reducing welding parameters ($\gamma = 7.5°$, $V_{con} = 1900$ m/s, $V_c = 250$ m/s), welding did not occur at all.

For the welded joint ($\mathbf{E}_w$): $\gamma = 15.2°$, $V_{con} = 2400$ m/s, $V_c = 634$ m/s. The arrangement of the plates is the same, the thickness of the plate of tantalum was 1 mm, aluminium 4 mm. Two steel substrates were used, the thickness of which was 5 mm and 15.5 mm. The impact was so strong that in addition to welding aluminium with tantalum, tantalum was welded to the upper steel plate.

The preparation of foils of dissimilar metal welds is associated with certain difficulties and using the traditional method of electropolishing it is very difficult to evenly thin them. Nevertheless, it was possible to prepare an Al–Ta foil using the following electrolyte: 10 parts of methanol, 6 – butyl alcohol, 1 – perchloric acid at a voltage of 30 V. In addition, foil samples were also obtained using the Fashione 1010 ION MILL ion cannon.

Figure 41 shows SEM images of the interface: flat for the welded joint ($\mathbf{E}_p$) (Fig. 41 *a*) and wavy for the welded joint ($\mathbf{E}_w$) (Fig. 41 *b*). For the wave-shaped interface, the period is approximately 300 μm, the amplitude 30 μm.

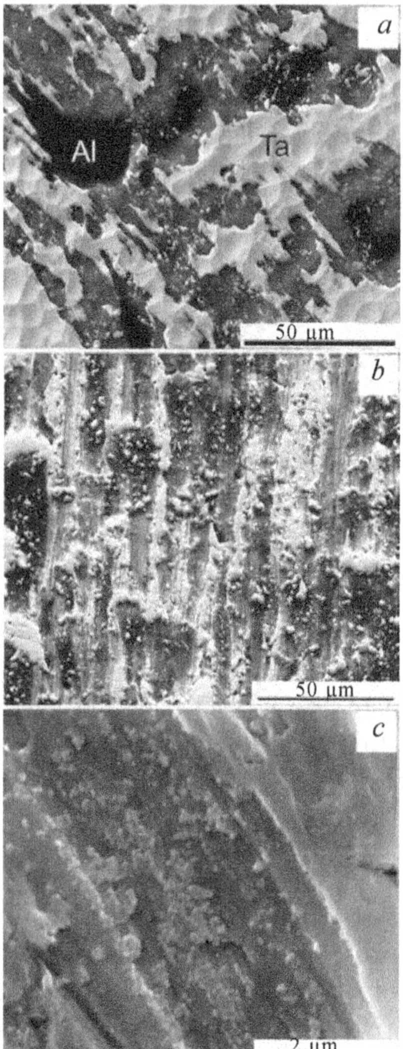

Fig. 42. SEM images of longitudinal section for compound ($\mathbf{E}_p$): *a* – the tricolour image; *b* – scattering of tantalum particles; *c* – aluminium etched.

Comparison of copper–tantalum and aluminium–tantalum welds allows us to clarify the question of the influence of the mutual solubility of the starting metals on such processes of the formation of welded joint as fragmentation, the formation of cusps and local melting zones. These zones are observed in welded joint ($\mathbf{C}_p$), ($\mathbf{C}_w$), ($\mathbf{E}_p$). For the welded joint ($\mathbf{E}_w$), a film is observed (frozen after melting) along the entire wavelike interface. The structure of the film turned out to be extremely important in clarifying the possible role of the melt for gluing the contacting surfaces. The results are published in [36, 38, 39], as well as in the reviews cited above [10, 12, 14].

3.3.1: (E_p) aluminium–tantalum welded joint, flat border

Figure 42 shows the SEM image of a longitudinal section of the transition zone for the welded joint (**E**$_p$). Here, as well as for the copper–tantalum welded joint (**C**$_p$), the transition zone consists of randomly distributed regions of three types (Fig. 42 *a*): tantalum, aluminium, and a gray zone containing both elements. This is a result of the appearance of cusps of one metal in another, namely tantalum into aluminium, as the hardest and densest metal in this pair. Using data on the chemical composition obtained by means of SEM, it was shown that the tantalum zone corresponds to white colour, the aluminium zone corresponds to black colour, and the mixture of initial metals to gray colour. In Fig. 42 *b*, a multitude of particles of tantalum of arbitrary shape against the background of aluminium is visible. We believe that the explosion of the particles of tantalum observed during explosive welding, as well as the above-

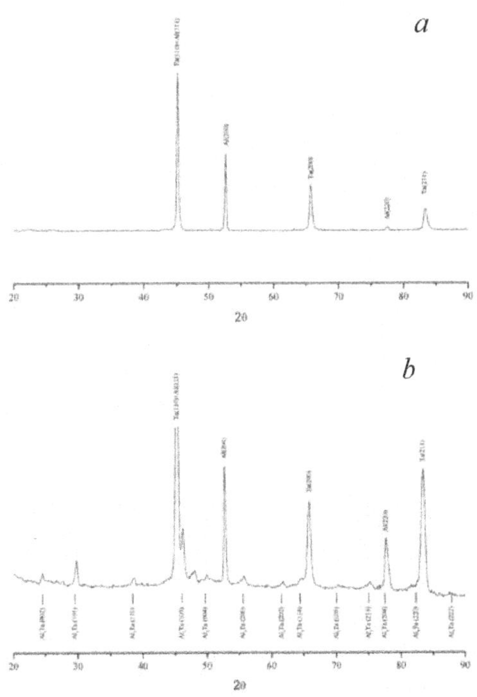

Fig. 43. The diffractogram of the transition zone (longitudinal section): *a* – welded joint (**E**$_p$); *b* – welded joint (**E**$_w$).

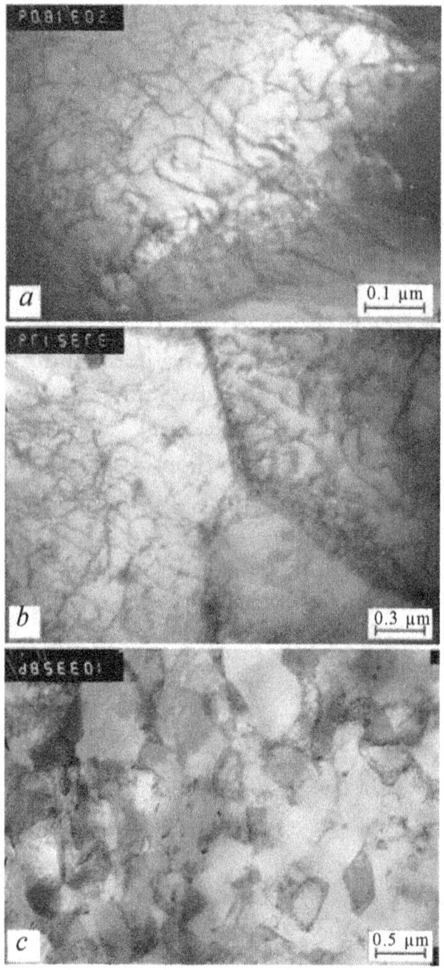

Fig. 44. Welded joint ($\mathbf{E}_p$) aluminium –tantalum, TEM image of the structure of highly deformed aluminium: cold hardening (*a*); recrystallized areas (*b, c*).

mentioned dispersion of aluminide particles for titanium aluminide welds and tantalum particles for copper–tantalum welds, is similar to the dispersion of fragments.

Figure 42 *a* shows the longitudinal section of the transition zone, obtained after the aluminium was completely etched. Tantalum particles are visible on the tantalum surface, which spill out during the etching of aluminium. In this regard, the image of tantalum particles in Fig. 42 is similar to that shown above in |Fig. 31 for the welded joint ($\mathbf{C}_p$). Particle scattering is observed both for metal–metal and metal–intermetallic welds, both in the presence and in the absence of mutual solubility, regardless of the shape of the interface (flat or wavy). This is proof of the dominant role of the explosion in the studied form of welding.

Earlier Fig. 30 showed colour micrographs of the longitudinal section for flat interfaces in the (**C**$_p$) and (**E**$_p$) welded joints. They are also similar, despite the different mutual solubilities of the starting materials. The observed tricolour images, as mentioned above, are the result of the occurrence of cusps on the interface by diffusion (due to the rapidity of welding) ejection of one metal to another. Only the formation of cusps can explain the picture of the transition zone for compounds (**C**$_p$) and (**E**$_p$) in Fig. 30. Of course, the topography of the interface for these welded joints is not identical. As can be seen from Fig. 30, local melting zones occupy a significantly larger area in the case of welded joint (**E**$_p$). Perhaps this is due to the fact that the melting point of aluminium (933 K) is lower than the melting point of copper (1381.6 K).

Since the Al and Ta metals have mutual solubility, as a result of explosive welding, the formation of a chemical compound should occur in the contact area of the metals. But the analysis of the transition zone of the welded joint (**E**$_p$) did not show the presence of intermetallic compounds: the diffraction pattern contains only the lines of tantalum and aluminium (Fig. 43 *a*).

The SEM results of point measurements of the chemical composition of the gray zone showed that the concentration of tantalum here does not exceed 5 at.%. The diffractogram of the transition zone for the welded joint (**E**$_w$) with a wave-like interface (Fig. 43 *b*) will be discussed in the next section.

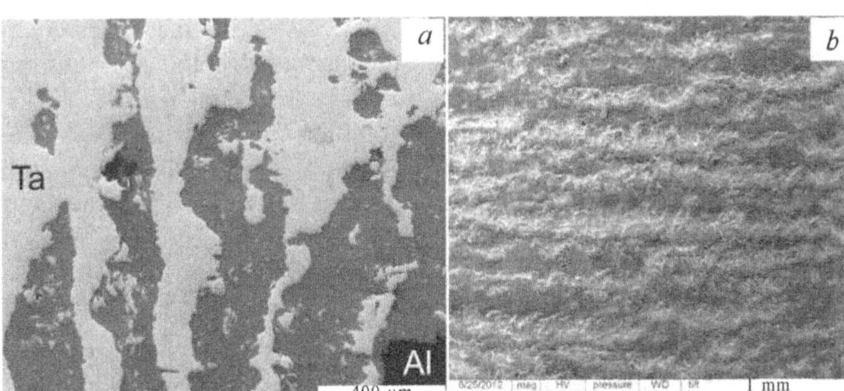

Fig. 45. The microstructure of the transition zone for the (**E**$_w$) welded joint, longitudinal section (SEM): *a* – band of aluminium and tantalum; *b* – aluminium is completely etched.

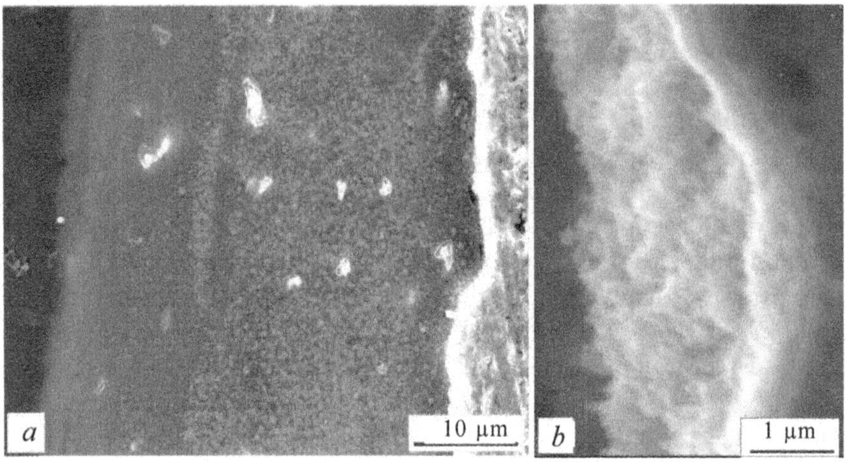

Fig. 46. Cross section of the transition zone for the ($\mathbf{E}_w$) welded joint: *a* – layers of molten aluminium with and without particles; *b* – spherulites.

Figure 44 shows for the aluminium–tantalum welded joint ($\mathbf{E}_p$) the TEM images of the strongly deformed, but not mixed with tantalum, aluminium structures: regions with high dislocation density (Fig. 44 *a*, recrystallized regions (Fig. 44 *b, c*)). Figure 44 *b* shows the formation of a triple junction. The images of the highly deformed structure of tantalum are similar to those observed for the copper–tantalum welded joint ($\mathbf{C}_p$). If we compare the images of the deformed structure of aluminium and copper (see Fig. 34), it is clear that aluminium has been recrystallized to a larger extent than copper. This is due to the fact that aluminium is, firstly, low-melting, and secondly, it has a higher stacking fault energy, which facilitates creep and cross-slip processes. This, in turn, contributes to the restructuring of the dislocation structure and, ultimately, recrystallization.

Although the temperature in the contact zone during explosive welding can be quite high, but with rapid explosive effects, the course of thermally activated processes that determine the movement and restructuring of dislocations is difficult and hardly possible. It can be assumed that these processes, like diffusion, become possible only at residual temperatures and stresses.

3.3.2. (Ew): aluminium–tantalum, wavy interface

The cross section for the ($\mathbf{E}_w$) welded joint is shown in Fig. 41 *b*. A longitudinal section (Fig. 45 *a*) is a set of alternating bands of tantalum and aluminium with almost parallel interfaces. In fact, this is a slightly inclined section, which is used because of the

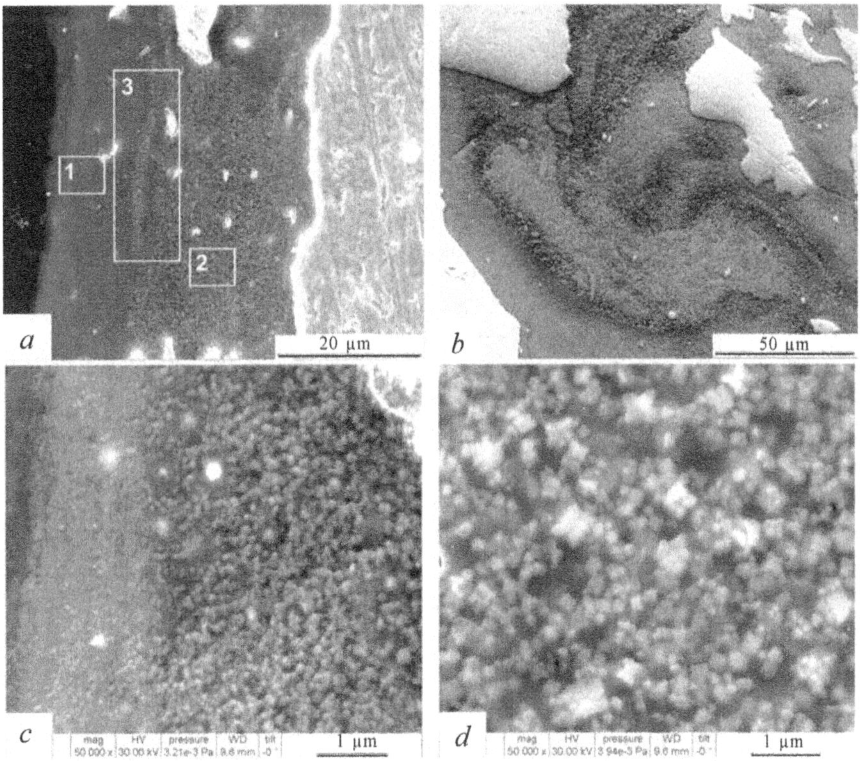

Fig. 47. The microstructure of the cross section of the transition zone for the welded joint ($\mathbf{E}_w$): *a* – layers of molten aluminium; *b* – vortex; *c, d* – areas 1 and 2 enlarged 20 times, marked in Fig. 47 *a*, respectively

small amplitude of the wavy surface. Figure 45 *b* shows the same longitudinal section, but after aluminium was etched. One can see the bands of tantalum, as well as particles, which spilled out when the aluminium dissolved. There is a clear similarity with the micrograph for the ($\mathbf{C}_w$) welded joint shown in Fig. 39

However, there is a significant difference in the structure of the transition zone for the ($\mathbf{E}_w$) and (**C**w) welded joints. Figure 37 *a* shows for the ($\mathbf{C}_w$) welded joint numerous local melting zones having a vortex-like structure. From Fig. 46 it immediately follows that, for the ($\mathbf{E}_w$) welded joint, in contrast to (C_w), does not contain local melting zones. Instead, in Fig. 46 *a*, if one moves from the solid phase of aluminium (left edge), layers of molten aluminium are visible, initially not containing and then containing particles of another phase (with a small, then with a high particle density). Particle sizes vary widely: from 50 nm to 500 nm. The rather sharp

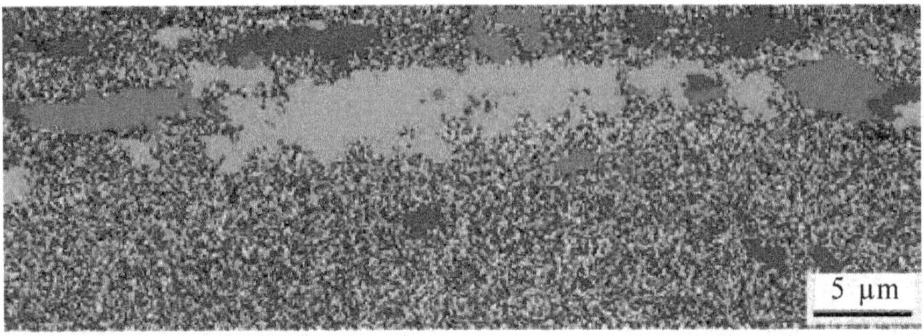

Fig. 48. (E_w) welded joint, the distribution map of the phases of aluminium, tantalum and intermetallic compounds, obtained by the EBSD method from section 3 of the mixing zone, shown in Fig. 47 *a*.

faceting of particles attracts attention. Next comes the edge of tantalum, observed in Fig. 46 *a, b*, like a glowing line. Such a luminous interface ('bundle'), which is a molten film, was observed earlier for the (A_p) titanium–aluminide welded joint (Fig. 23 *b*). Thus, the film includes a tantalum edge with a width of 2...2.5 μm and the above-mentioned molten layer, which has a width of about 40 μm (30 μm with particles, 10 μm – without particles).

As can be seen from Fig. 43 *b*, on the diffractogram, taken from the longitudinal surface of the thin section, X-ray peaks are clearly recorded not only from tantalum and aluminium, but from the Al_3Ta intermetallic phase. The Al_3Ta compound has a tetragonal crystal lattice, an I4/mmm space group, and an Al_3Ti structural type. Obviously, the melting of tantalum is required for an intermetallic reaction. Indeed, in Fig. 46 *b* shows spherulites which are witnesses of the melting of tantalum. A similar structure has the above-mentioned molten film for the titanium–aluminide welded joint (A_p) where spherulites were also observed (see Fig. 23 *e*).

Figure 47 shows the microstructure of the transition region in more detail. From Fig. 47 *c, d*, which are magnified images of areas 1 and 2, and accordingly, it is clear that as the distance from unmelted aluminium moves away, particles in the melt layer increase and acquire a rather sharp faceting, which was not observed in the (E_p) welded joint. In the titanium–VTI-1 and Cu–Ta welded joints, vortex formation was observed in the local melting zones. Here, in the (E_w) welded joint, there are no local zones, but vortex formation is also observed inside the molten layer (Fig. 47 *b*). The formation of the intermetallic phase Al_3Ta was observed in the melting zone of the Al–Ta (E_w) welded joint. The clusters preceding the formation of the

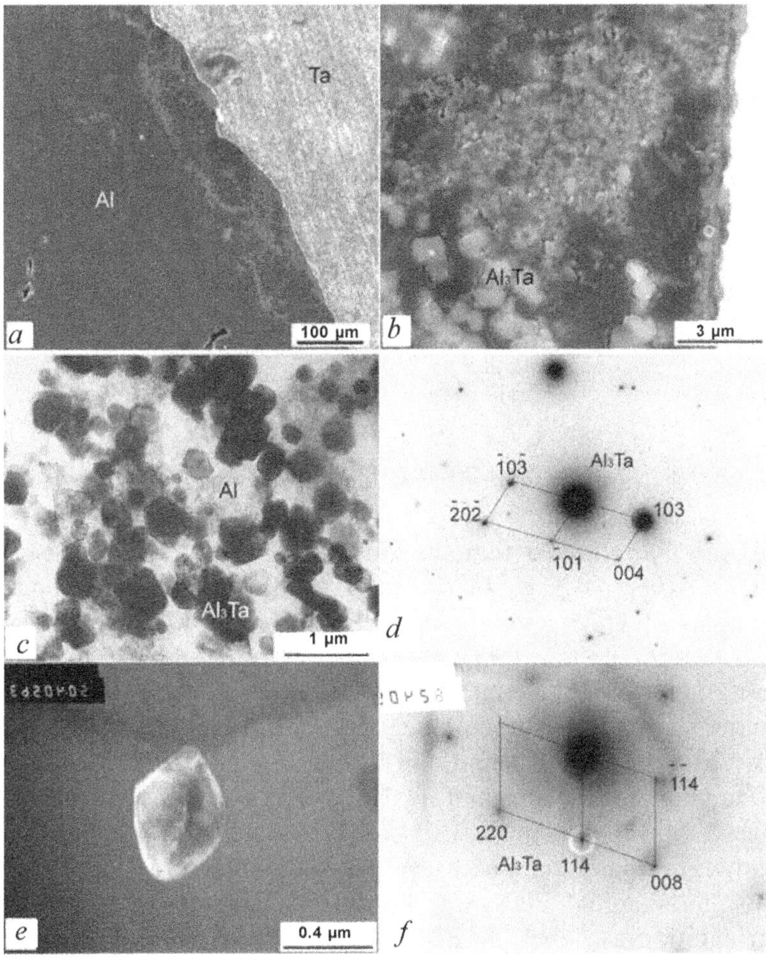

Fig. 49. Welded joint (**E**w) after annealing at 500° C: *a, b* – SEM images of the cross section at different magnifications; *c* – TEM image of the Al region with intermetallic inclusions; *d* – microdiffraction from the site shown in Fig. 49 *c*, zoe axis [010]; *e* – intermetallic particle in aluminium in the reflex (11$\bar{4}$) (dark field); *f* – microdiffraction from a site presented on Fig. 49 d, zone axis [$\bar{1}$10].

Al_3Ta intermetallic compound are visible in Fig. 47 *c*, and particles Al_3Ta – on Fig. 47 *d*. Using SEM, EBSD analysis of the structure of the transition zone of the ($\mathbf{E}_w$) welded joint was conducted in the region marked with figure 3 in Fig. 47 *a*. Figure 48 is a map of the distribution of the various phases of the Al–Ta system in the mixing zone. With the help of EBSD, the morphology of the phases

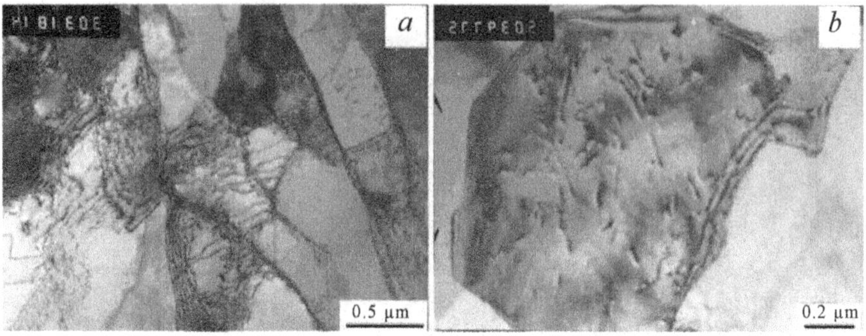

Fig. 50. Traditional fragmentation of Al for the ($\mathbf{E}_w$) welded joint: *a* – cellular and banded structure; *b* – recrystallized grains.

in a frozen aluminium melt was detected. First of all, attention is drawn to the heterogeneity of the distribution of phases. The EBSD method showed that the amount of the Al_3Ta intermetallic phase is ~50%. In Fig. 48 the intermetallic phase correspond to large areas of green colour. Monochrome areas of blue colour are fragments of tantalum, taken off to the mixing area. The total share of tantalum was approximately 30%. A structure representing an ultramicrocrystalline (three-colour) mixture of phases (Al, Al_3Ta and Ta) is observed on the remaining image area (Fig. 48).

.Computer processing and analysis of misorientations using the software attached to the EBSD prefix showed that the interfacial and intergranular boundaries are predominantly medium and high angle. The share of low-angle borders is less than 3%. The mean misorientation angle of the EBSD analysis was approximately 42°.

To study the growth dynamics of the intermetallic phase, we performed annealing at 500°C for 1 h, followed by cooling in water. Figure 49 *a, b* is an SEM image of the structure of a layer containing intermetallic inclusions after annealing. It is seen that the structure is inhomogeneous, but on the whole, the process of formation of intermetallic compounds and enlargement of particles was completed, as compared with those observed in Fig. 47 *c*. The TEM study of the annealed samples confirmed the formation of the Al_3Ta intermetallic phase. In Fig. 49 the particles are well distinguished, which, as shown by the decoding of the electron diffraction pattern presented in Fig. 49 *d*, are intermetallic Al_3Ta compound. It was possible to obtain images and microdiffraction for individual particles of the intermetallic compound (Fig. 49 *e, f*).

Far from the transition zone, the unmelted aluminium for the welded joint ($\mathbf{E}_w$), contains structures characteristic of the

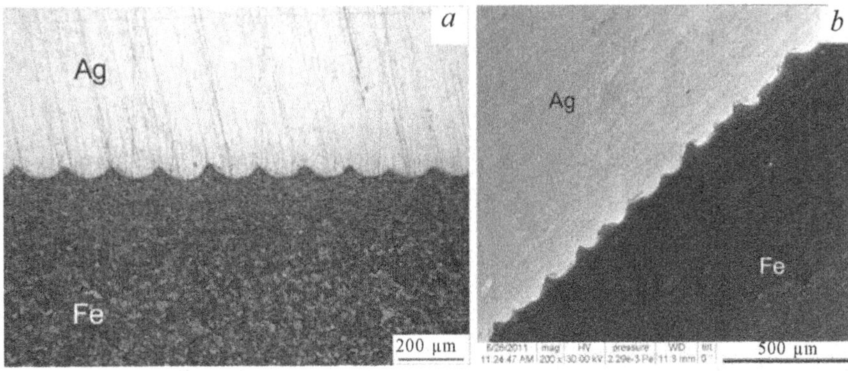

Fig. 51. Fe-Ag (Dw) welded joint, wavy interface (cross section): *a* – optical micrograph; image; *b* – SEM image.

fragmentation occurring during severe deformation (Fig. 50). These are cellular and banded structures (Fig. 50 *a*), as well as recrystallized grains (Fig. 50 *b*). Images of the highly deformed structure of tantalum are similar to those observed for the copper–tantalum welded joints (Fig. 34). As a result of annealing, aluminium grains increased about 100 times, but a very high dislocation density remained. Obviously, there was a collective recrystallization during annealing.

The structure of the transition zone (Figs. 46, 47) observed for the ($\mathbf{E}_w$) welded joint is formed as follows: first melting aluminium as a more low-melting metal, then melting the tantalum surface, forming a solid solution based on aluminium, on reaching certain concentrations forming particles of the intermetallic phase. However, before the melting begins in accordance with the scenario proposed above, the most rapid process occurs: the dispersion of solid particles of tantalum.

The fact that the melted region for the ($\mathbf{E}_w$) welded joint is a film at the interface, while in the ($\mathbf{E}_p$) welded joint there are isolated zones, is consistent with the intensification of the welding mode used to produce the ($\mathbf{E}_w$) welded joint compared to the compound ($\mathbf{E}_p$). This is also related to the observation of intermetallic inclusions only for the ($\mathbf{E}_w$) welded joint.

For the previously studied titanium–aluminide, copper–tantalum, and aluminium–tantalum welded joints, both a flat interface was observed with some welding modes and a wave-like one with others. For the iron–silver welded joint, we limited ourselves to the study of only the wave-shaped interface and a comparison with the

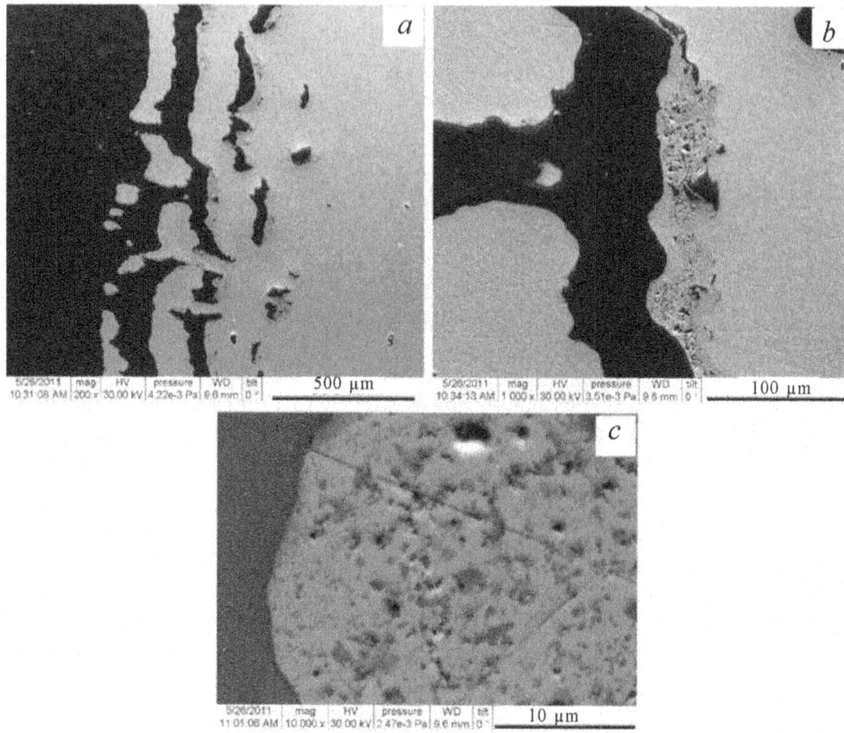

Fig. 52. Fe–Ag ($\mathbf{D}_w$) welded joint, the heterogeneity of the interface (longitudinal section): *a* – cusps, *b, c* – zones of local melting (with different magnification).

Table 7. The results of spot measurements of the chemical composition of the local melting zone

Measurement No.	Fe content, at.%	Ag content, at.%
1	97	3
2	97	3
3	0	100
4	56	44
5	62	38
6	100	0
7	2	98

copper-tantalum wave-like welded joint. Both pairs of materials have practically no mutual solubility. The results are published in [14, 40], as well as in the reviews cited above [10, 12, 13].

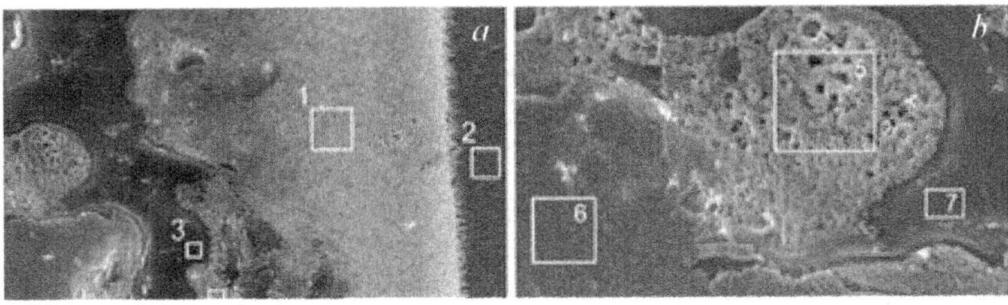

Fig. 53. The Fe–Ag ($\mathbf{D}_w$) welded joint, local melting zone (longitudinal section): *a* – near the interface, *b* – the internal structure of the zone, indicated measurement points.

To obtain the iron–silver welded joint ($\mathbf{D}_w$), a parallel arrangement of plates was used. The fixed plate was located on a metal substrate. Welding was performed at the following parameters: $\gamma = 15.6°$, $V_{con} = 1910$ m/s, $V_c = 520$ m/s, $W_2 = 0.73$ MJ/m^2.

Here, W_2 is the energy expended on plastic deformation. Armco iron plate thickness was 1.5 mm, silver plate 2 mm.

Figure 51 shows the optical micrograph and SEM images of the iron–silver interface. The wavy shape of the interface is fairly regular (Fig. 51 *a*), although it contains waves of different amplitudes and even almost straight sections (Fig. 51 *b*). On average, the wavelength is about 140...150 μm and the amplitude 50 μm.

For a wavy boundary, the longitudinal section is a set of alternating bands of iron and silver with parallel interfaces. The SEM image in Fig. 52 was obtained not with longitudinal but with an inclined section (white strips are silver). The size of the cusps is tens of microns, reaching in some cases 100 μm. As numerous SEM measurements of the cusp composition have shown, these are iron cusps. As can be seen from Fig. 52 *a*, the strips can break off and fall apart. At the same time, in some areas, the wavy nature of the interface is lost. Local melting zones (darker than adjacent silver strips) are already visible in Fig. 52 *a*, and at higher magnification in Fig. 52 *b*. These zones are clearly visible in Fig. 52 *c*: dark particles of iron against a background of silver.

Data on the chemical composition of the local melting zones (Fig. 53 *a, b*), obtained using SEM according to numerous measurements, are shown in Table 7.

As can be seen from Table 7, the chemical composition in the center of the zones is about 60Ag–40Fe (at.%). Figure 53 shows the areas of measurement. In Fig. 53 *b* iron particles are clearly visible

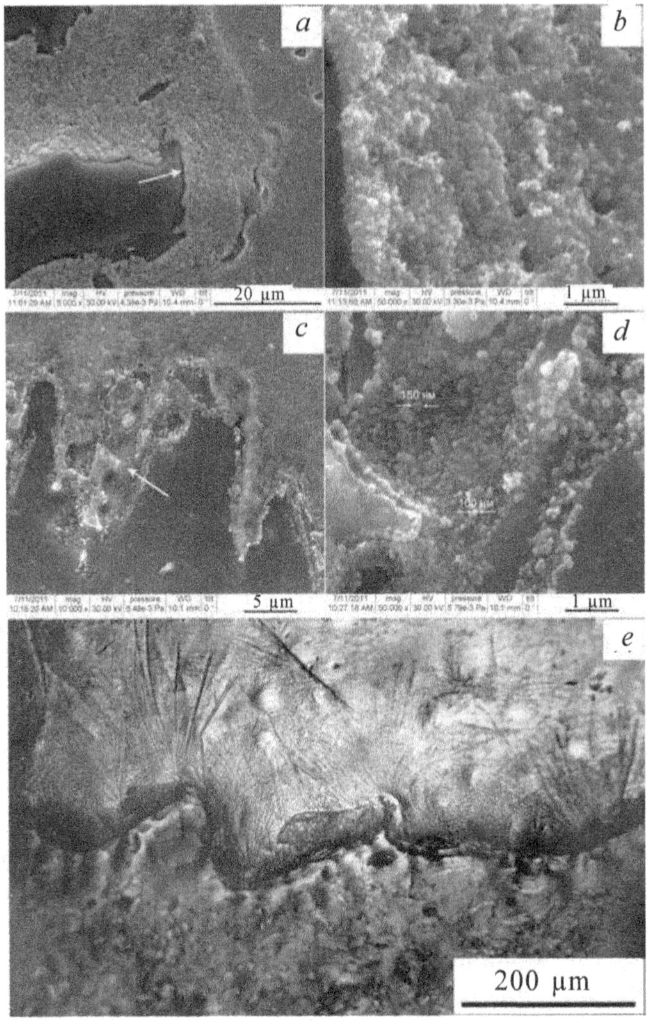

Fig. 54. The Fe–Ag welded joint (D_w), local melting zone: longitudinal section, *a, c* – at lower magnification; *b, d* – at high magnification, silver spherulites; *e* - cross section.

on a silver background. It is significant that the structure of the local melting zone is inhomogeneous: below the square with the number 5, iron particles are almost invisible. This means that there are both concentrated and non-concentrated solutions of iron in silver.

The internal structure of the local melting zones is shown in Fig. 54. In Fig. 54 *a* the arrow marks the area, the structure of which with a larger magnification is shown in Fig. 54 *b*. A similar structure

Fig. 55. Fe–Ag (Dw) welded joint, longitudinal section of the transition zone (iron etched): silver crests visible.

is also shown in Figs. 54 *c, d.* As can be seen from Fig. 54 *b, d,* spherulites formed during the hardening of the silver melt,spherulites: ultrafast hardening, many nuclei, but their growth is inhibited. Attention is drawn to their rather perfect spherical shape, which provides a minimum of surface energy. The spherulites have sizes of 100...200 nm.

Previously, in various titanium–aluminide welded joints (different aluminide compositions, different modes), spherulites were observed only for the titanium–VTI-1 alloy welded joint ($\mathbf{A}_p$) and only when the flat interface was molten. Figure 23 *e* is the SEM image of the structure of such an interface containing titanium spherulites. When compared with Fig. 54 *d*, it can be seen that the size of titanium spherulites (200...400 nm) is larger than that of the silver spherulites. For the copper–tantalum $\mathbf{C}_p$ and $\mathbf{C}_w$ welded joints, spherulites were not observed. The image of the tantalum spherulites for the ($\mathbf{E}_w$) aluminium–tantalum welded joint is shown in Fig. 46 *b*.

In the iron–silver welded joint the local melting zone consists of silver containing iron particles. At the same time, unlike the titanium–aluminide welded joint (Fig. 13) no vortices were observed. Vortices, but imperfect, were observed for the ($\mathbf{C}_w$) welded joint (Fig. 37 *b*).

Fig. 56. Fe–Ag (D_w) welded joint, TEM image of the local melting zone (iron is etched): the voids remaining from the iron particles are visible.

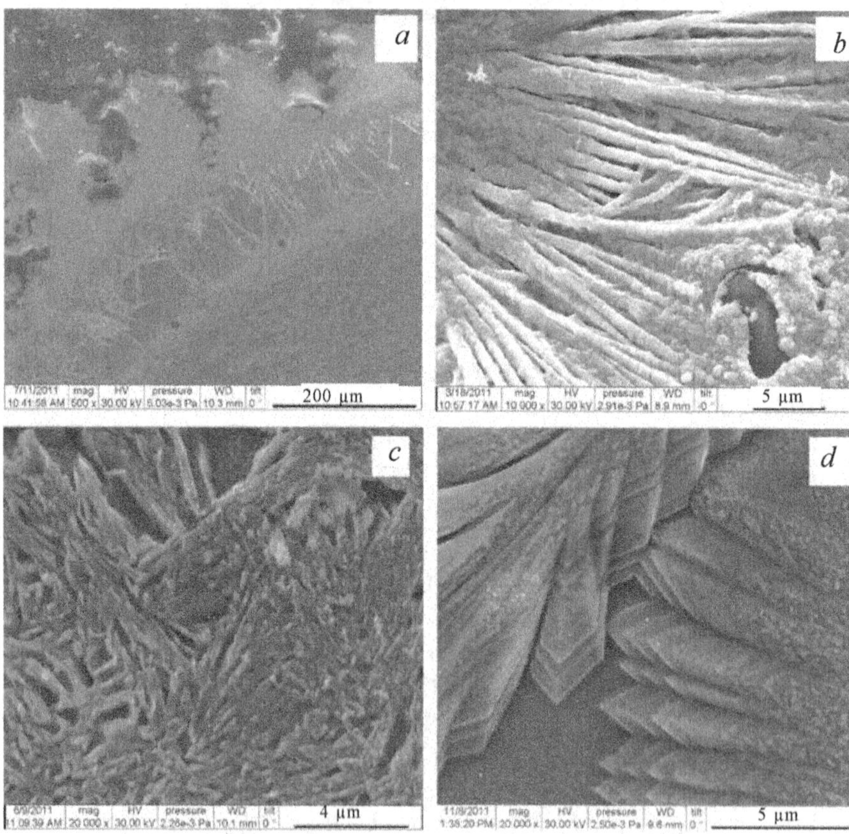

Fig. 57. Fe–Ag (D_w) welded joint, silver areas filled with dendrites, (SEM): a–d – at different magnifications.

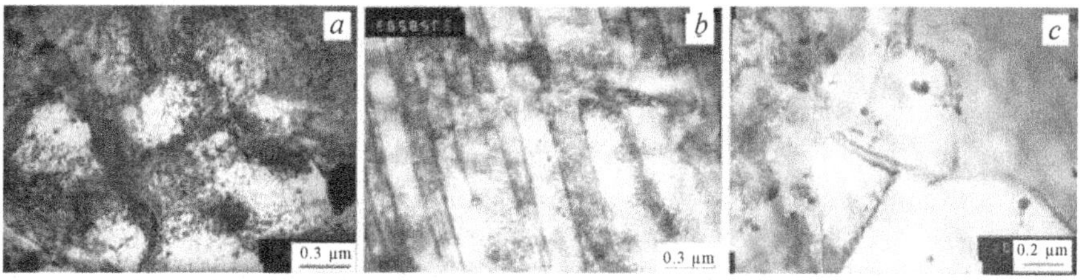

Fig. 58. Fe–Ag (D_w) welded joint, structure of highly deformed silver: *a* - cellular; *b* – band, *c* – recrystallized.

The imperfection of the vortices, as mentioned above, may be due to the increased viscosity of the melt containing solid particles. But it remains unclear why no vortices were found in the iron–silver welded joint. As can be seen from Fig. 54 *e* (cross section), along the entire interface, including the local melting zone. The area filled with dendrites is stretched. The width of the region (vertically from the interface) is approximately 200...250 µm.

Some of these features of the interface are clearly visible in a more prominent SEM image obtained after the iron was completely etched (Fig. 55). Instead of bands, Fig. 55 shows silver crests which are broken lines. A constriction between them, significantly distorting the interface, is also clearly visible. Observed in the longitudinal section of the violation in the alternation of bands, including the loss of continuity, reflect the imperfection of the interface.

In the chapters 7 and 8, a similar surface observed in the Cu–Ta and Cu–Ti welded joints is called quasi-wave. It looks like a patchwork quilt. The similarity of the images of the surface relief of silver (iron etched) shown in Fig. 55 with those shown in Fig. 118 *b* and 120 *b* images of the relief of the surface of titanium (copper etched) is clearly visible. In all these cases, the observed relief is a strip-type patchwork.

Figure 56 shows the TEM image of the local melting zone. One can see black voids left by the iron particles when they were etched. Judging by the observed voids, the iron particles had an arbitrary shape. Electron diffraction shows only the presence of silver.

Returning to Fig. 53 *a*, we note that the entire area on which dot 1 consists of close-packed dendrites, which, as is well known, are witnesses to melting. Separate dendrites germinate in the neighbouring region of silver where there was no melting. The indicated area with a different magnification is visible in Fig. 57.

Dendrites are visible, with both primary and secondary branches. Both dendrites and a fine crystalline interdendritic region are visible.

Note the similarity of the images shown in Figs. 57 and 26: cusps similar to 'fingers' in Fig. 26 *b* and 57 *a*, dendrites and the interdendritic region in Fig. 26 g and 57, c.

Figure 58 shows cellular (*a*), band (*b*) and recrystallized (*c*) silver structures, which are typical for fragmentation under severe deformation. We believe that the observed structures refer to a solid body that has not undergone melting. Similar structures were observed for all welded joints examined above.

As is known, dendritic growth begins when there is a negative temperature gradient in the liquid phase.This is indeed the case, since the most 'hot' is the area near the interface. But if melting occurs almost instantaneously, during an explosion, the growth of dendrites occurs already at residual temperatures, when diffusion becomes possible.

The thermal conductivity coefficient of silver is about 418.7 W/m·deg and is one of the highest for metals. The thermal conductivity coefficient for iron is 74.4 W/m·deg. The difference in the coefficients leads to the fact that the heat sink from the interface is carried out through silver. The thickness of the molten silver layer is about 200 μm. Below, a calculation will be made to find out if the energy supplied is sufficient to melt the said silver band. The following parameter values for silver will be used: atomic weight 107.9, density 10.5 g/cm^3, melting point T_{melt} = 1234 K, melting heat Q_{melt} = 11.34 kJ/mol, heat capacity $C_v \approx 24$ J/(mol K). From here it is not difficult to show that the energy required for melting of the specified layer is approximately $E_{\text{melt}} \approx 0.23$ MJ/m^2, and for its heating by 1000 K $E_{\text{heat}} \approx 0.48$ MJ/m^2. Thus, the total energy required for heating and melting a layer of silver with a thickness of 200 μm is $E_{\text{tot}} \approx 0.71$ MJ/m^2.

The estimate obtained is overestimated, since all the energy pumped during the explosion for the iron-silver system is only 0.73 MJ/m^2, but the calculation did not take into account the necessary energy consumption for fragmentation, heating of iron, heating of silver away from the melting region, etc. However, the fact that the layer containing dendrites has a thickness less than 200 μm in many places was not taken into account, and this can lower the value of the energy required for its melting. Thus, in general, it can be assumed that the energy balance will be sufficient to heat and melt the silver layer.

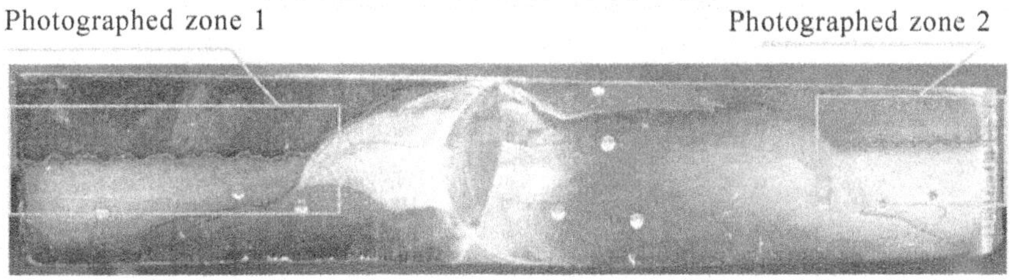

Fig. 59. Cross-section of the steel 08Kh13–steel 12KhM welded joint (Sw) near the joining interface.

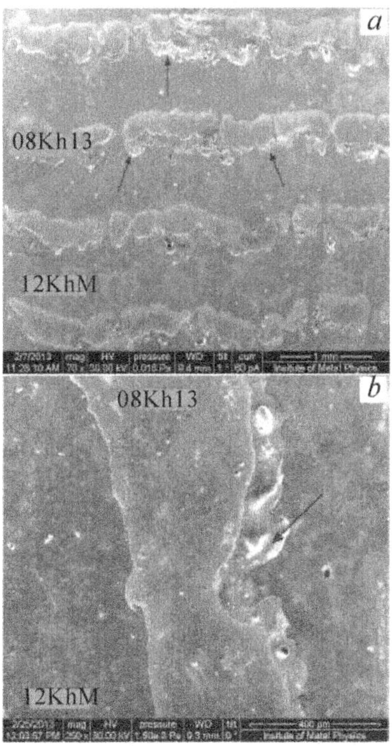

Fig. 60. Steel 08Kh13–steel 12KhM welded joint (S_w), wavy interface, longitudinal section; *a*, *b* – zones of local melting (indicated by arrows) at different magnifications.

The microhardness of the iron–silver welded joint was measured at various points in the transition zone. It was found that microhardness values (about 750 MPa) in the zone of local melting of silver containing iron particles exceed the smallest microhardness value obtained for a silver region with a dendritic structure (approximately 450 MPa). However, this excess is incomparably smaller than that found earlier for the zone of local melting of copper containing tantalum particles. It can be assumed that the stronger dispersion hardening observed in the case of the copper–tantalum welded joint is associated, firstly, with the greater hardness of tantalum compared

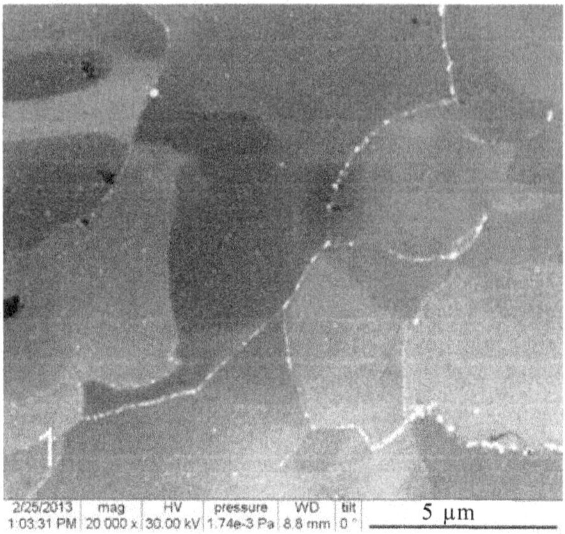

Fig. 61. Zone 1 – recrystallized steel 12KhM.

to iron, and secondly, with a higher concentration of the tantalum particles compared to the iron particles.

3.5. Steel–steel

The steel 08Kh13–steel 12KhM composite produced by explosive welding was used in the manufacture of a shell of a coke oven chamber. Further, the structure of the welded joint (**S**$_w$) of these steels [41] is investigated in order to identify risk zones for the shell of the coke oven chamber, discussed in Section 5.2.

Coke oven chambers are hollow vertical cylindrical devices with a diameter of 3...8 m and a height of 22...40 m with the mode of operation in conditions of sufficiently active media. The chamber walls investigated in the present work were made of a two-layer composite. For the inner shell of the chamber, chromium steel 08Kh13 was used, which is corrosion- and heat-resistant steel (up to 750...800°C), and the outer shell was made from a low-alloyed low-carbon steel 12KhM.

Table 8. Chromium concentration at measurement points

Point	*a*	*b*	*c*	*d*	*e*	*f*	*g*
Cr, wt%	1.28	4.81	4.59	4.12	9.35	4.40	7.80

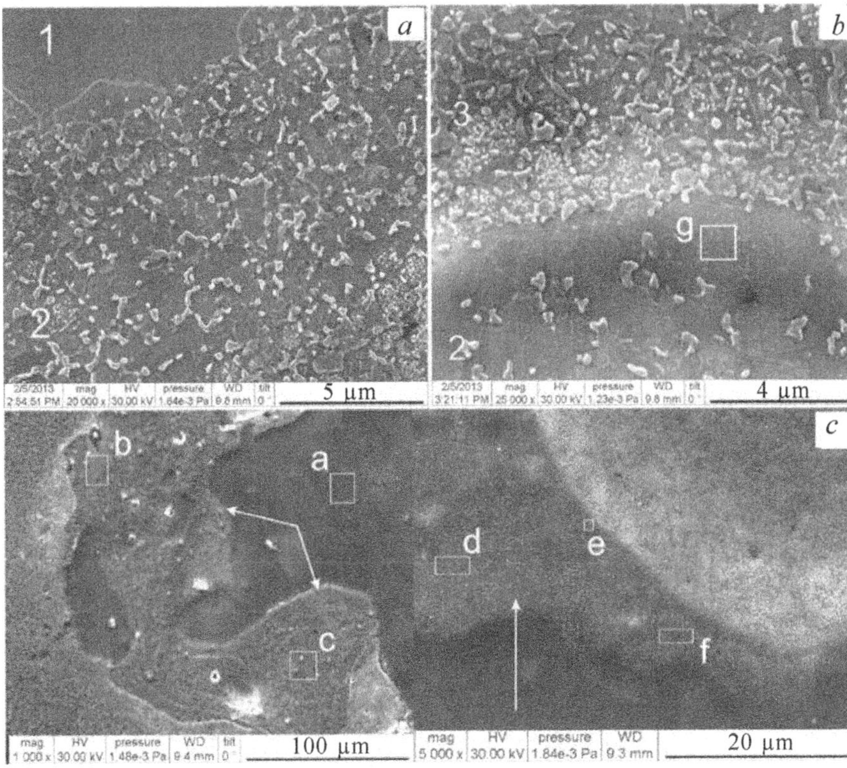

Fig. 62. Welded joint (S_w), transitions between zones: *a* – zones 1, 2 (2 – melt); *b* – zones 2, 3; *c* – measurement points in the melt zone (indicated by arrow).

The sheets obtained by explosive welding have a length of 4...5 m and a width of up to 2.4 m. The plate thickness from 08Kh13 steel was 6 mm, the plate thickness from steel 12KhM was 24 mm, the gap between the plates was 5 mm. The velocity of the contact point was 2400...2700 m/s. The collision of the plates occurred at an angle of 9...11°. After welding, tempering was carried out to relieve residual stresses at 680...700°C for about 2 hours.

Figure 59 shows the cross-section of a welded joint near the joining of double-layer sheets. Visible wave-like interface on opposite sides of the dock.

Figure 60 shows the longitudinal sections of the wavy interface, which are alternating bands of both steels. At different magnifications isolated molten areas are visible, i.e. zones of local melting (indicated by arrow). As can be seen from Fig. 60, melting zones are not observed inside the bands, but only near their borders, and they are pressed to the border from the side of steel 12KhM. As a result of

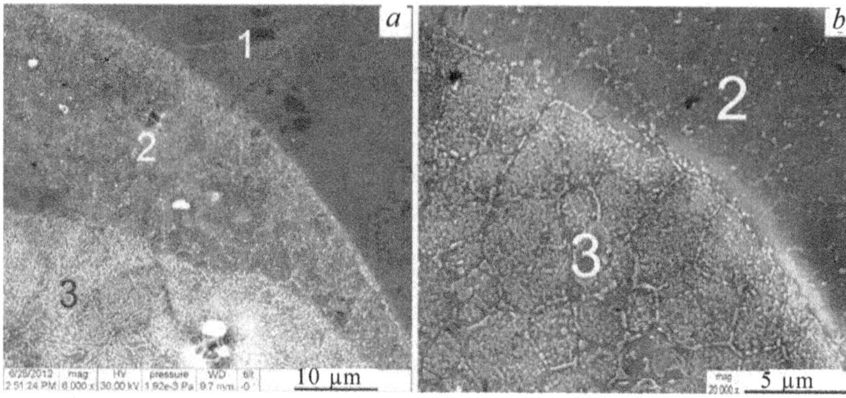

Fig. 63. Welded joint (S_w), zone 3 and adjacent zones: *a* – zones 1, 2, 3; *b* – sharp boundary between zones 2 and 3.

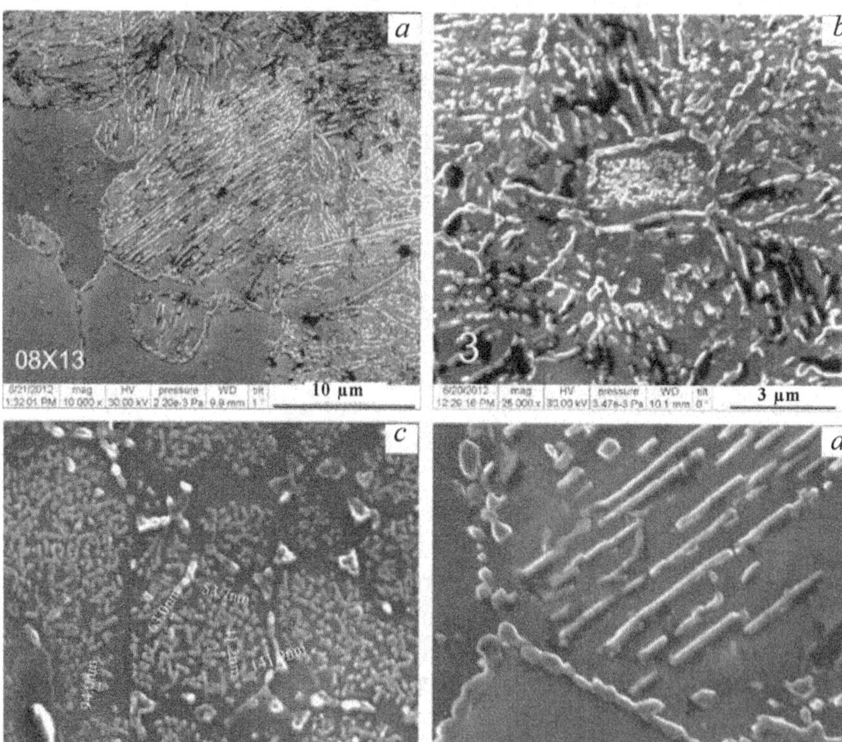

Fig. 64. Welded joint (S_w), zone 3 – colonies of rod-shaped carbides: *a*...*d* – at different magnification.

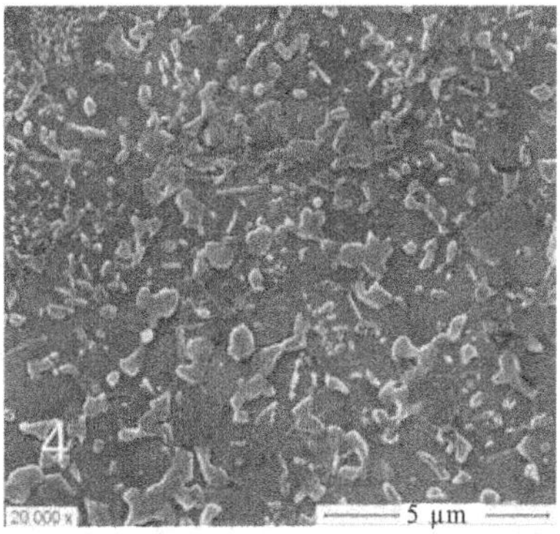

Fig. 65. Welded joint (S_w), zone 4 – steel 08Kh13, carbides of arbitrary shape, no faceting.

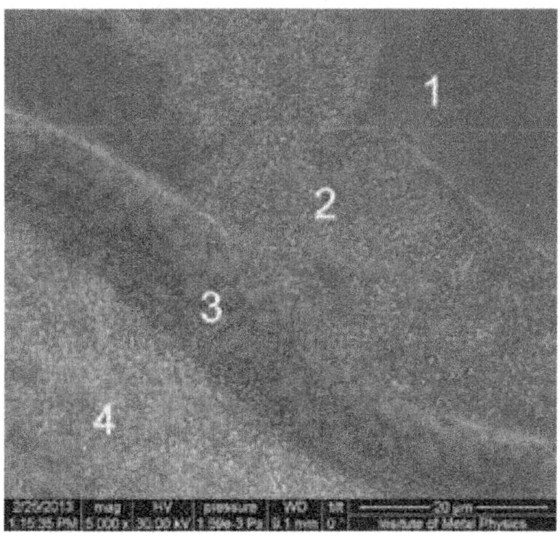

Fig. 66. Transitions between zones in the steel.

the SEM analysis, it was shown that the structure of this welded joint consists of layers and the transitions between these layers are investigated further, starting from steel 12KhM, then through the interface, the transition to steel 08Kh13. For convenience, the layers are numbered.

Figure 61 is an SEM image of the recrystallized 12KhM steel area (area 1). Rare inclusions of cementite are visible along the grain boundaries. Area 1, as can be seen from Fig. 62 *a, b*, borders with the local melting region 2, the images of which with a lower magnification are shown earlier in Fig. 60. As shown by chemical analysis carried out by SEM, the melt is chromium-rich and carbon-lean steel 12KhM. Table 8 shows the chromium concentration

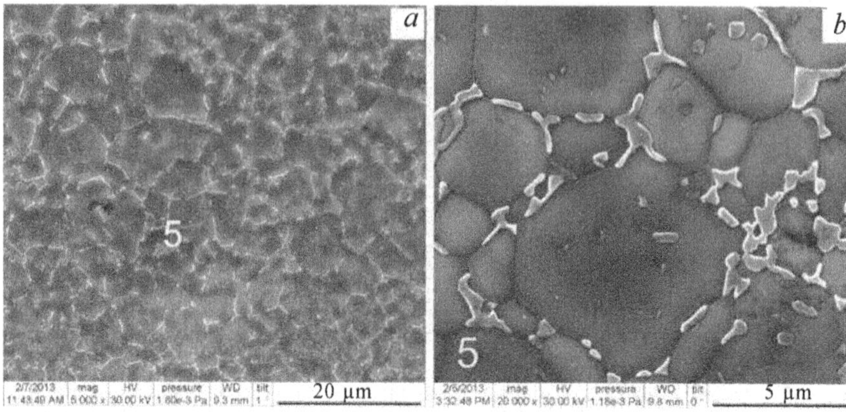

Fig. 67. Welded joint (C_w), zone 5 – recrystallized steel 08Kh13: *a, b* – with different magnification.

values at the measurement points. For comparison, the values of the chromium concentration at the point *a*, which is in region 1, are also given. The values of the chromium concentration at the *b*–*g* points of the melt are several times higher than the chromium concentration at the point *a*. As shown by numerous measurements (some values are given in Table 8), the melt is characterized by a random distribution of concentration inhomogeneities. Correspondingly, carbides of various types are observed that are characteristic of both low- and high-chromium steels.

As can be seen from Figs. 62 *b* and 63, region 2 is in contact with region 3, in which the density of carbides is incomparably higher. Area 3 is located along the edge of steel 08Kh13. As can be seen from Fig. 63 *a*, there are so many carbides that zone 3 appears white. The sharp border between areas 2 and 3 in Fig. 63 *b* is clearly visible.

Figure 64 *a* shows a colony consisting of thin rods of various lengths lying in the section plane. Some of them are located so close that they almost merge, others – at much greater distances. Here and in Fig. 64 *b* colonies of a different orientation are visible.

The rod sections are irregular polygons, indicating the cutting of rods. The colonies have a different density of carbides: high density for the central colony in Fig. 64 *b*, significantly smaller for the neighbouring colonies. As shown in Fig. 64 *c, d*, the thickness of the rods is approximately 40...60 nm. At the grain boundaries, large carbides of another type are visible, with a thickness of 100...150 nm.

Area 3 is adjoined by area 4 of steel 08Kh13, containing carbides of arbitrary shape, without faceting (Fig. 65).

Figure 66 shows the sequence of four layers: 1 – recrystallized region of steel 12KhM; 2 – zone of local melting of steel 12XM; 3 – zone of rod-like carbides in steel 08Kh13; 4 – zone of randomly distributed (and possibly different structure) carbides in steel 08Kh13.

And again the sharp border between the narrow region 3 and the wide region 2 draws attention. Such a border contrasts with the blurred border between the regions 3 and 4. The sequence of layers completes layer 5 of recrystallized steel 08Kh13 (Fig. 67).

In Fig. 67, numerous carbides are visible both at the boundaries and inside the grains. As can be seen from the comparison of Figs. 61 and 67, in the recrystallized steel 08Kh13 the density of carbides is higher, and the grain sizes are smaller than in the recrystallized steel 12KhM.

Measurements of microhardness in various bands of the transition zone showed that the microhardness of the rod-like carbide zone (zone 3) reaches a value of 3600 MPa, which is about 2000 MPa higher than the microhardness of recrystallized steel (region 1, 5). Such a high value is due not only to the microhardness of the carbides, but also to the fact that they are combined into colonies. As a result of the appearance of colonies, the deformation behaviour of the material, due to the action of significant and long-term loads, may change significantly, including, possibly, in the direction of reducing the long-term strength. Indeed, during operation the walls of the reactor chamber are subjected to pressure due to the mixing of

4

Discussion of results

4.1. Fragmentation of the granulating type

The phenomenon of fragmentation has long been known. It is observed in biology, genetics, medicine, geology, and in many other areas. The simplest example: fragmentation of algae and primitive worms as a way of their reproduction. The most beautiful example: the fragmentation of coral reefs. Numerous variants of cell fragmentation with which a breakthrough in modern genetic engineering is associated are widely known.

Explosive fragmentation is the process of separation (crushing, granulating) of a solid into parts (fragments) that occurs under a strong external influence. The theory of particle dispersion in an explosion was created by Neville Mott [42] during the Second World War. Accordingly, the names of the weap""on: *fragmentation warhead, fragmentation shell, fragmentation bomb.*

Mott, together with other authors (see, for example, Grady's book [43]), showed that, using simple geometric fragmentation methods, which consist of dividing straight lines into separate segments, one can describe the dynamic fragmentation of a cylindrical shell. Figure 68 *a* shows a line randomly broken into fragments with variable length l and average length λ. Then the number of fragments $N(l)$ having a length greater than l is equal to

$$N(l) = N_0 \exp(-l/\lambda) \tag{1}$$

For the number of fragments having a mass greater than m, a similar exponential dependence was obtained, but only on m. In a one-dimensional model, the fragmentation of a straight line imitated the fragmentation of a glass tube or stretched wire for which the experimental data were described by a few more complex

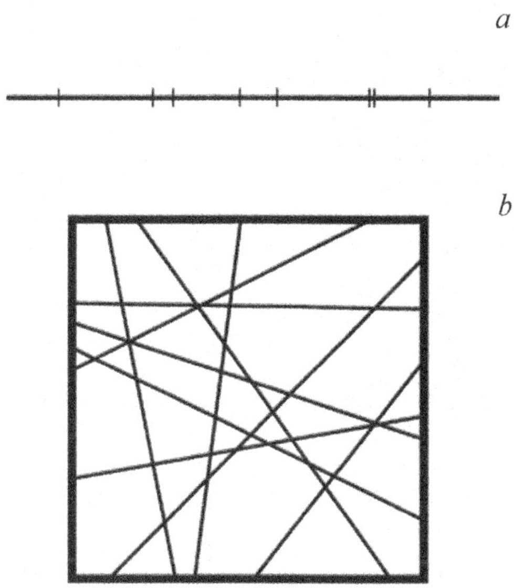

Fig. 68. Examples of geometric chaotic fragmentation: *a* – a straight line, randomly divided into segments of different lengths; *b* – a random set of straight lines to describe the splitting of the surface.

distributions. But in order to get closer to the topology of the cylinder, other variants of geometric fragmentation were used, which are sets of arbitrarily oriented straight lines. One of such sets is shown in Fig. 68 *b*. As a result, various fragments distributions were obtained that describe the distribution of fragments, including the different shapes of the expansion fields. The subsequent development of the approach developed by Mott, in particular, the inclusion of statistical heterogeneity, is highlighted in the Grady monograph mentioned above [43].

We believe that the dispersion of fragments exploded by Mott and the dispersion of particles during explosive welding have much in common. As can be seen from Fig. 18 (compound ($\mathbf{A}_w$) titanium - orthorhombic titanium aluminide), the dispersion of particles is similar to the dispersion of fragments in an explosion, only other sizes. But during an explosion the fragments fly apart in open space, while in explosive welding, particles scatter in the closed space between the plates. In Fig. 18, the aluminide region is easily distinguished due to the α_2-phase globules. It is seen that it is from this area that particles of arbitrary shape fly out, which, naturally, are

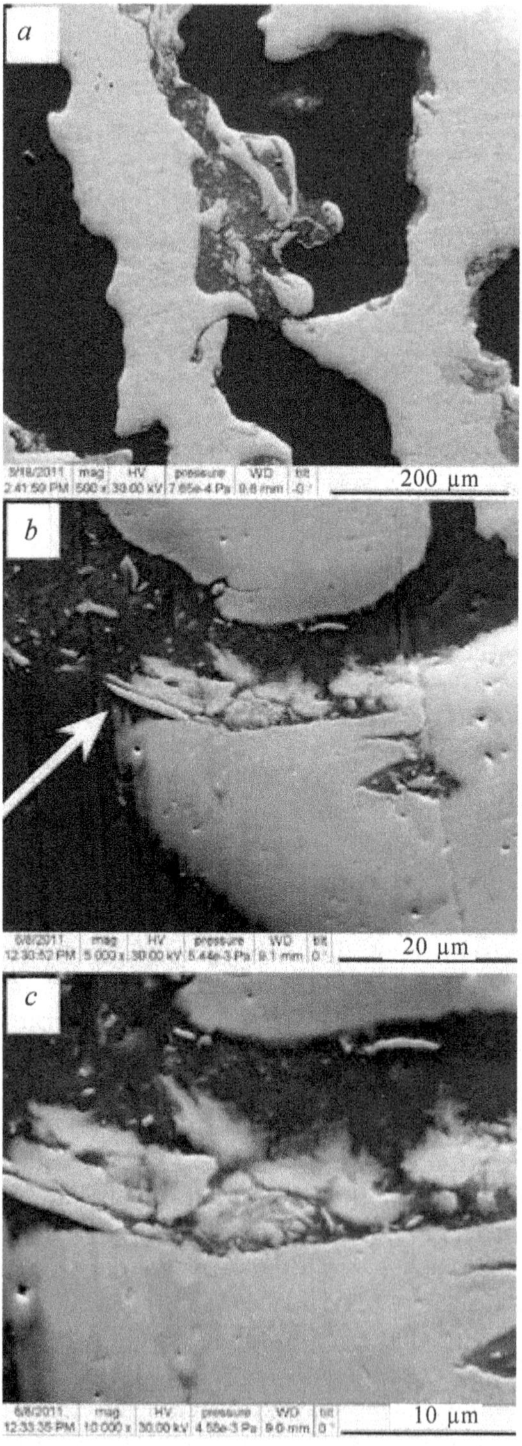

Fig. 69. Dispersion of tantalum particles near a large cusp for the compound (C_w) copper - tantalum: *a* – a large cusp; *b, c* – fragmented layer at different magnifications.

aluminide particles. This leaves a fragmented layer that consists of aluminide particles that are partially in contact with each other. This layer is particularly visible in Fig. 18 *c*. There are many particles inside the local melting zones. These are aluminide nanoparticles in titanium (Fig. 17 *d*), tantalum nanoparticles in copper ((**C**$_p$), Figs. 32, 33), micron tantalum particles in copper ((**C**$_w$), Fig. 38 *b*), micron particles of iron in silver ((**D**$_w$), Fig. 53 *b*). It is these particles that pour out from the local melting zones, clearly visible against the background of tantalum, when copper ((**C**$_p$), Fig. 31) or aluminium ((**E**$_p$), Fig. 42 *c*) are etched.

Figures 69 and 70 illustrate the scenario of how the formation of particles, the expansion of some and the partial consolidation of others. We draw attention to the large cusp of tantalum (approximately 100 microns), marked by an arrow. In the nearest copper strip, a zone of local melting is visible, containing numerous tantalum particles. The remaining particles of tantalum, which did not have time to fly away, are clearly visible in Fig. 69 *b, c*.

Figure 70 (**A**$_w$) titanium–aluminide welded joint) shows the formation of a strip consisting of aluminide particles. The particles are close together, but not very tightly. Submicron particles of arbitrary shape fly out to the neighbouring areas of local melting (Fig. 70 *b, c*).

Many details of the scattering of particles during explosive welding and fragments during an explosion are common. In the monograph [44] it is noted that in the spectrum of the fragments there are two dissimilar morphological aggregates: large fragments and accompanying small fragments. Similarly, in explosive welding, as can be seen from Fig. 69, and, in addition to many small particles of tantalum, filling the melting zone, large particles of tantalum are observed.

In addition, in [44], an image of the fragments on the contact

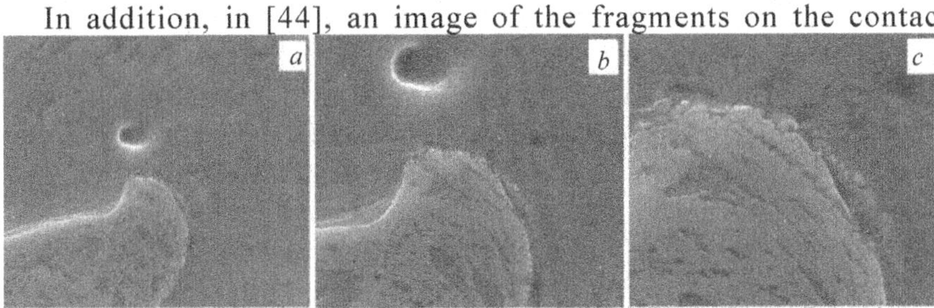

Fig. 70. Aluminide particles near a large cusp for titanium-aluminide welded joints:*a* - at low magnification; *b, c* – at high magnification, visible particles at the top of the cusp.

2). Similar features of the dispersion of particles during explosive welding are shown in Figs. 69 and 70, on which the particles are also visible that have not had time to separate from the contact surface.

In [44], the stages of fragmentation during an explosion were formulated: the nucleation of cracks, the growth of cracks, their fusion and the formation of fragments. Similar stages of formation and scattering of particles during explosive welding were considered by us in [23, 26]. They are the stages of the fragmentation process observed in explosive welding, for which we proposed the term: fragmentation by crushing (FBC). Perhaps this is not quite a good process name. More compact could be the name: fragmentation fragmentation. But in English, because of the coincidence of words - a fragment and a fragment - such a name would not make sense. In this book we use the term: *granulating fragmentation*.

Granulating fragmentation (GRF) is a process of separation into particles, which either scatter or merge with each other. In other words, the GRF includes both the expansion of particles and their partial consolidation. GRF is an analogue of explosion fragmentation. In both cases, there is an expansion of particles (fragments), but only with the GRF, the continuity of materials is preserved. Granulating fragmentation is reviewed in [12]. GRF was observed for all welded joints studied by us. The results are presented in the relevant sections.

A deformable body, being an energy-non-equilibrium system, tends to include the most efficient channels of energy dissipation. Under the severe conditions that are realized during explosive welding, including the absence of relaxation mechanisms for dislocations, one of the most effective channels for energy dissipation can be free surfaces arising during the propagation of microcracks. An important point to which attention was drawn in [45] is the following: at very fast loads, plastic deformation can be expected to be delayed in relation to stress for several microseconds. Therefore, taking into account the high loading rate during explosive welding and the high rate of this process, we believe that the dissipative channel, which is turned on before others, is the formation of elastic microcracks. Microcracks are considered to be discontinuities with dimensions smaller than those of Griffith [45]. When there is a set of microcracks, their size and shape are preserved until $\sigma < \sigma_{cr}$, where $\sigma_{cr} = \sqrt{\frac{2\gamma E}{\pi l_{max}}}$.. Here, σ_{cr} is the critical stress, l_{max} is the maximum crack size, γ is the specific energy of the free surface, E is Young's modulus. If the stress does not change, then at a size smaller than the critical one, it is advantageous to arrest the crack, and if it is larger, the crack will grow.

Unlike the fragmentation process investigated by Mott during an explosion, which leads to fracture, GRF occurs as a result of microfracture and is an alternative process to fracture. Instead of free surfaces, the occurrence of which would lead to fracture, the formation of surfaces associated with microfracture occurs, which either belong to scattering particles or 'heal' during their consolidation. Microcracks do not arise under the action of medium applied stresses, but on their local stress concentrators. It is in areas containing local stress raisers that multiple microcracks can occur. Appropriate free surfaces experience inhibiting when they meet each other. First, it prevents the transformation of microcracks into macrocracks. Secondly, microvolumes arise, surrounded on all sides by free surfaces. These microvolumes are actually cut from the surrounding material. Numerous SEM observations of particle expansion have been given above. Fragmentation as a division of material into weakly connected microvolumes is an alternative process to fracture. GRF increases the 'vitality' of the material, saving it from fracture even with such a strong external effect, which is explosive welding. This is the essential role of the GRF in explosive welding.

GRF was studied by us for various welded joints. The GRF is observed both for metal–metal welds and metal–intermetallic compounds, both in the presence and in the absence of mutual solubility of the initial metals, regardless of the shape of the interface (flat or wavy). Among the set of studied compounds there was not one where GRF would not be observed. This means that the GRF is a universal phenomenon inherent in explosive welding. It is the presence of an explosion that makes fragments inevitable. GRF is the most rapid process during explosive welding, which manages to occur during the explosion.

We assume a single scenario in which the coupling of the GRF with other processes is carried out: an explosion, the formation of particles (fragments), local dispersion of particles, local melting, and preservation of a layer of non-dispersed particles. Further, in section 4.4, the relationship between the scattering of particles and local melting is discussed.

4.2. Fragmentation under severe deformation

In explosive welding, apart from the GRF, one more type of fragmentation is observed. This refers to fragmentation, the existence of which is confirmed by numerous observations of the structure of various materials subjected to severe deformation. This refers to SPD by torsion, equal channel pressing, rolling, forging, as well as combinations of these methods [46–49]. This, to a certain extent, traditional fragmentation, includes the pumping of dislocations and twins, the formation of tangle, cellular and band structures, nanocrystallization, recrystallization.

Although the temperature in the area of contact during explosive welding can be quite high, but during such a stage, the thermally activated processes that determine the movement and restructuring of dislocations can not take plac. It can be assumed that these processes, like diffusion, become possible only at residual temperatures and stresses. It is these processes that, in contrast to the SPD, determine traditional fragmentation. Both types of fragmentation occur at different distances from the contact surface: GRF in a narrow region near the surface, while traditional fragmentation is somewhat farther from the surface.

Both types of fragmentation were observed in each of the welded joints investigated by us. Typical microstructures for traditional fragmentation can be seen in the following figures: Fig. 12 ($\mathbf{A}_w$), Fig. 25 ($\mathbf{A}_p$), Fig. 34 ($\mathbf{C}_p$), Fig. 44 ($\mathbf{E}_p$), Fig. 50 (E_w), Fig. 58 ($\mathbf{D}_w$). Here are the names of the welded joints. The results are also presented in the above-mentioned review [12] devoted to both types of fragmentation. The similarity of microstructures both with each other and with known microstructures obtained earlier under severe deformation attracts attention. The specificity of traditional fragmentation during explosive welding is that structures typical for various stages of developed plastic deformation are observed for welded joints at the same time.

We will continue the comparison of explosive welding with severe deformation of solid material. If we confine ourselves to torsion under pressure (SPD by torsion), then the values of pressure and energy input are close to those used in explosive welding. But the temperature and the duration of deformation differ significantly: torsion under pressure is carried out at room temperature or at 77 K, within minutes or even tens of minutes. Of the two types of fragmentation, only the PDT is to some extent similar to the transformations of the structure under SPD by torsion, including its refinement.

4.3. Consolidation of powders with SPD by torsion

A strong external effect on a substance has various goals: a radical change in its structure, which provides optimization of the properties, expanding the field of materials use, including initially brittle ones, increasing the possibility of using materials in harsh environments. Such an impact can be achieved by various methods such as SPD (severe plastic deformation by torsion), ECAP (equal channel angular pressing), rolling, forging, as well as combinations of these methods. In addition, it is impossible not to mention explosive welding, in which pressures are used (of the order of several GPa), which are close, for example, to those with SPD by torsion.

The nanostructured materials obtained as a result of a strong external effect cause anomalously high interest in comparison with other materials because of the unique complex of their inherent properties. Many of them are of a substantial and immediate practical interest. It can also be noted that identifying the relationship of the structure with the properties of such substances is an important fundamental problem.

The SPD by torsion method can be used not only for grinding the structure, but also for consolidating powders. During deformation by torsion at high pressure, the bulk of the material in the form of a disk experiences hydrostatic compression caused by pressure as well as pressure from the outer layers of the sample. The consequence of this may be the preservation of the continuity of the material even at high degrees of deformation. As examples, here one can cite a strong compaction using SPD by torsion of metallic powders (micron or nanocrystalline) of Cu, Al, Ni, Ti, as well as their compositions [49, 50].

The main question in the analysis of the transformation of powder into a solid material is as follows: why does powder particles stick together? For metal powders, the natural mechanism is the joint plastic deformation of the contacting particles, accompanied by the restructuring of the dislocation structure. It was of interest to expand the field of research and move from materials with a metal interatomic bond to materials with other types of bonds.

One of the interesting systems in this regard semiconductor materials with a covalent bond.

It is known that the plastic flow in semiconductors begins only at sufficiently high temperatures [51]. Ultimately, this means that, in a wide temperature range, not only dislocations are blocked, but, more importantly, dislocation sources as well. Therefore, there is no reason to assume the possibility of generating dislocations in such materials under torsion under pressure. However, the nanostructuring of silicon was discovered in [52]. We believe that the indicated structural transformation occurs without the participation of dislocations and is carried out precisely by means of the GRF: the powder is ground and at the same time docking of newly formed particles occurs.

To confirm this view, experiments were performed [12], in which the initial silicon powder with a sufficiently large and uniform particle size of about 64–100 μm was used at different modes of SPD by torsion. Parameter values (pressure, angle of rotation) were found at which a sufficiently smooth plate appeared from the initial powder. It was shown that immediately the plate does not form after pressing and at small angles of rotation. The microstructure of silicon – powder or plate – is shown in Fig. 71. It was found that the resulting plate was very fragile, i.e. particle adhesion is weak. Nevertheless, the observed strong crushing of the initial powder and its subsequent consolidation bring together the structural transformation under SPD by torsion with fragmentation of the

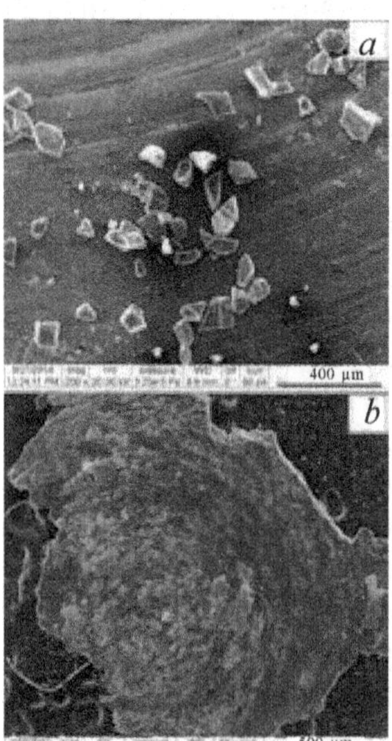

Fig. 71. Microstructure of silicon after powder consolidation by torsion under pressure (9 GPa, 0.25 rotaions): *a* – the initial powder, *b* – the general view of a thin plate [12].

granulating type during explosive welding. However, the question of the mechanism of consolidation has not yet a definite solution. It is not clear whether the consolidation is carried out as a result of deformation by means of dislocations and twins, or the covalent nature of interatomic bonds determines a different mechanism of sticking of particles.

Therefore, in order to find out the possibility of consolidation in the absence of dislocations, electron microscopic studies of a number of essentially fragile materials subjected to SPD by torsion were carried out [53]. The following crystalline materials were selected: quartz and crystal (intermediate type of bonds). In addition, glass (object and quartz glass) having an amorphous structure (Van der Waals forces) was investigated.

Deformational consolidation of powders was carried out by high-pressure torsion in Bridgman anvils. The anvil pistons are made of VK-6 superhard alloy, the diameter of the working sites of the pistons is 4 mm, respectively, the consolidated samples had the same diameter. The average thickness of the samples was 120...130 microns. The rotational speed of the anvil was 0.3 rev.min. The pressures and angles of rotation for each of the materials are presented in Table 9.

The study of the initial and consolidated states was performed on a Quanta 200 Pegasus scanning electron microscope. The phase composition was studied by X-ray diffraction analysis, taking samples on a DRON-3 X-ray diffractometer in monochromatic copper radiation with a step of 0.05° and counting at 10 s.

Table 9. Powder consolidation modes

Powder material (SiO_2)	Treatment pressure P, GPa	Angle of rotation φ, deg	Treatment conditions
Slide	10	10	I
	12	15	IIa
Quartz tube glass	10	10	I
	12	10	II
Quartz for calibration of GOST 9077-82 diffractometer	10	10	I
	14	10	II
Rhinestone	14	10	II

In the initial state, the studied materials is a system of isolated particles. If two surfaces are in contact, then taking into account their inherent roughness, it can be assumed that contact occurs first on the tops of the surface asperities [54, 55]. The relief teeth can wrinkle and flow into depressions, be cut off, or shred and crush into crumbs, which in some cases can fill a crack. Small irregularities while grind. Over time, the number of particle contacts increases. As a result, consolidation occurs and the solid body contains (or does not contain) pores, contains (or does not contain) microcracks.

The dominant factor is adhesion, which in turn depends on the crystal structure, crystallographic orientation, surface cleanliness, normal load, temperature, contact duration, interatomic interaction forces, etc. But because of the misorientation mentioned above, the lattice mismatch may occur when they are joined, which leads to an increase in surface energy.

In the metal crystals, the presence of dislocations and twins causes the adjustment of the lattices, which contributes to the consolidation. Below are the results of an experimental study of the consolidation of non-metallic materials by torsional SPD..

4.3.1. Quartz

The crystal structure of quartz (frame type), is built of silicon–oxygen tetrahedra, arranged in a spiral. Quartz has a fairly high degree of polymorphism. At low temperatures, the lattice of crystalline quartz, built on the basis of SiO_2, has a trigonal system (α-quartz), and in the temperature range 573...870°C – hexagonal (β-quartz). The melting point is approximately 1713...1728°C [56].

Quartz is a brittle material at any temperature and pressure, and its fracture, as a rule, occurs by cleavage. Unlike silicon, which has covalent interatomic bonds, in quartz there is an intermediate type of bonds: covalent bonds, supplemented by ionic bonds.

Quartz has many varieties: rock crystal, smoky crystal, amethyst, etc. Rock crystal, which is a colourless transparent variety of quartz, was also investigated by us with SPD by torsion.

In addition to crystalline quartz, amorphous quartz can be obtained from cooling of molten quartz. It is the quartz glass that has various uses. The results of the study of quartz glass with SPD by torsion are given below in the appropriate section.

Figure 72 *a, b* shows schematically the structures of crystalline quartz and glassy quartz. Since the figure shows a diagram in a

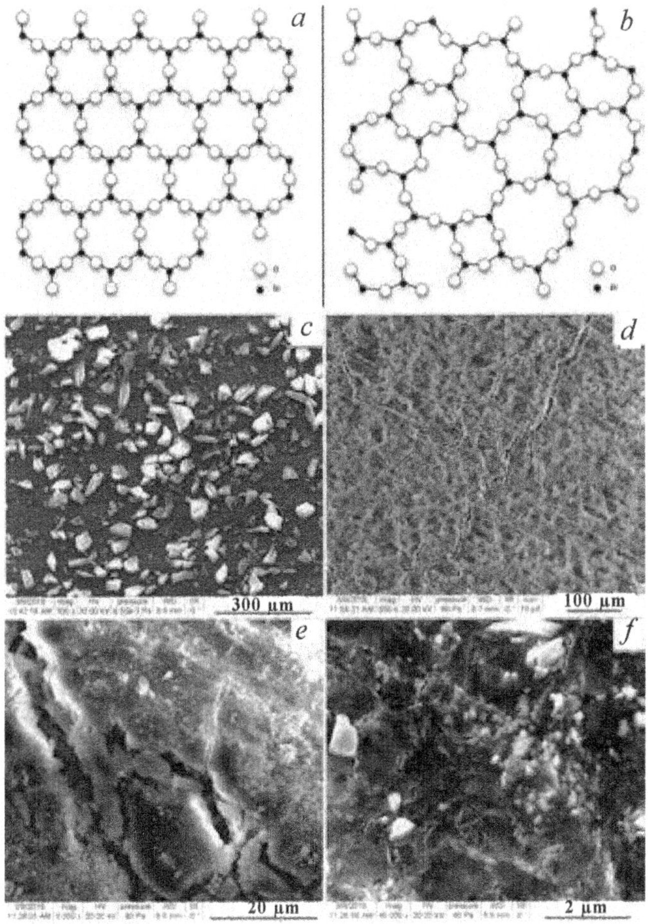

Fig. 72. Schematic representation of the structure of crystalline (*a*) and glassy quartz (*b*); the structure of the initial quartz powder (*c*), fragmentation of particles and their partial consolidation (e...fe) as a result of pressing at a pressure of 14 GPa.

two-dimensional image, each silicon atom is surrounded here only by three, and not four, oxygen atoms.

In the study of quartz, a plate of single-crystal α-quartz was used, hereinafter referred to as the initial plate. Similar plates are used to set up X-ray diffractometers.

The SEM image of the powder obtained by grinding the initial plate is shown in Fig. 72 *c*. The article size varies in the range 50...150 μm.

The subsequent pressing of the initial powder (in Bridgman anvils) was carried out at a pressure of 14 GPa, but without torsion. At the

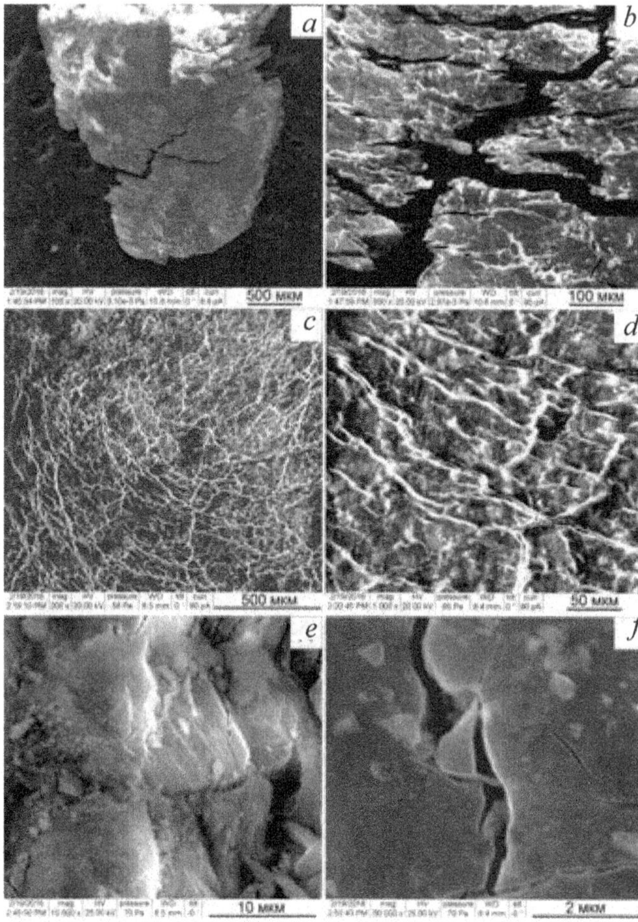

Fig. 73. The structure of the surface of quartz after SPD in torsion (pressure 14 GPa, rotated by 10°): *a* – plate; *b* – edges of the macrocrack; *c, d* – network of microcracks with different magnification; *e, f* – fracture.

same time, fragmentation of particles and their partial consolidation were found (Fig. 72 *d*). As a result, there is not a solid plate, but a plate consisting of separate parts and at the same time small particles (Fig. 72 *e, f*).

In addition, SPD was performed at a pressure of 14 GPa, with rotation though 10°. As a result, a plate was obtained and its SEM image is shown in Fig. 73 *a*. The internal structure of the macrocrack with a larger magnification is visible in Fig. 73 *b*. Here one can see that the shores of the macrocrack are heavily cut.

The dense network of microcracks is shown in Fig. 73 *c, d* at different magnifications. In the absence of plastic deformation, precisely such a network of microcracks filling the entire plate is a channel for stress relaxation at such a high level of energy input, which is realized in SPD with torsion. It can be assumed that a significant discharge of the supplied energy falls precisely on the free surfaces belonging to microcracks.

In quartz, among microcracks one can see especially long ones that cross a significant part of the pattern. It is significant that even they do not turn into macrocracks. The typical features of the branching of microcracks are also clearly seen [54]: configurations of the triple joint type, which are clearly visible in Fig. 73, c. In addition, the arcs are visible, corresponding to the rotation of the anvil, to which individual particles adhere.

At a sufficiently large magnification, one can see that the microcrack consists of straight sections (Fig. 73 *d*). It can be assumed that they are located in the cleaved planes.

Figure 73, *d, e* shows cracks on the surface of the fractured plate, obtained by SPD by torsion. It is of interest to observe in certain places of the crumbs that form during crushing and partial filling of microcracks with it, which is also discussed in [54]. The particle inside the crack is clearly visible in Fig. 73 *f* at a high enough magnification.

According to X-ray structural analysis, the results of which are presented in Section 4.3.3, the initial α-phase of quartz is retained during powder consolidation.

4.3.2. Rock crystal

Rock crystal is a colourless, transparent, usually chemically pure, almost without impurities, type of low-temperature modification of quartz SiO_2 crystallized in a trigonal syngony. It is found in the form of solitary or collected in druzes of crystals. Pure, defect-free rock crystal crystals are relatively rare and highly valued. In addition, it is possible to artificially obtain rock crystal.

The SEM image of the original crystal powder is shown in Fig. 74 *a*. Particles having a size of about 100...300 μm are visible, as well as many smaller particles. As a result of SPD by torsion, a plate was obtained (Fig. 74 *b*), which contains a network of microcracks, as well as macrocracks with rugged edges (Fig. 74 *c*). In Fig. 74 *d*, and with a larger increase in Fig. 74 *e*, the features of the microcrack branching mentioned above are clearly visible.

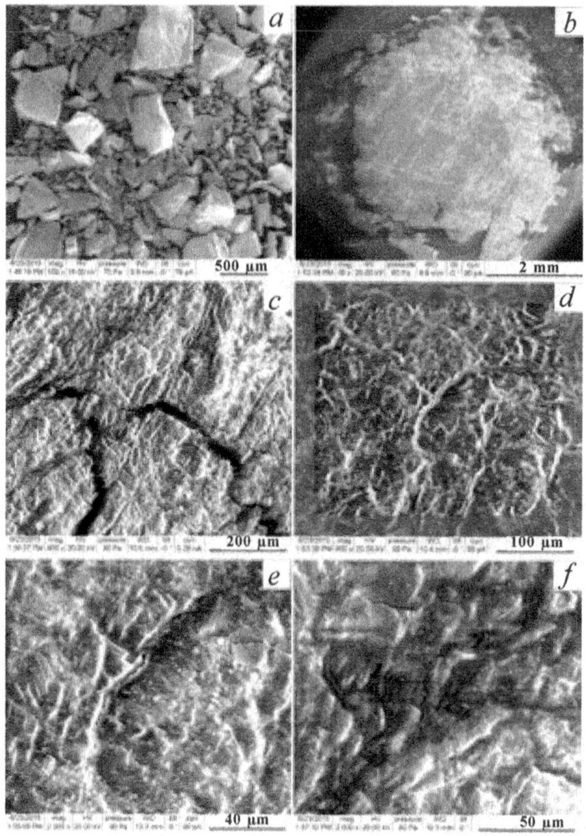

Fig. 74. The structure of the surface of rock crystal after SPD by torsion (pressure 14 GPa, rotated through 10°): *a* – the original powder; *b* – plate; *c* – grid of microcracks and macrocrack; *d, e* – branching of microcracks; *f* – particle formation and crack filling.

Branch witnesses are characteristic triple-junction configurations, as discussed above. All these configurations of microcracks are similar to those observed for quartz (Fig. 73 *c*). Particularly noteworthy is the similarity of the images in Fig. 74 *f* for crystal and Fig. 73 *e* for quartz. In both cases, the filling of microcracks with a fine powder is seen.

4.3.3. X-ray analysis

Figure 75 shows diffractograms for crystalline quartz, and Fig. 76 – for glass-like quartz given in the upper half of Fig. 75, the fragment of the diffractogram of the initial powder corresponds to the crystal modification of α-quartz. Below in the same figure is a

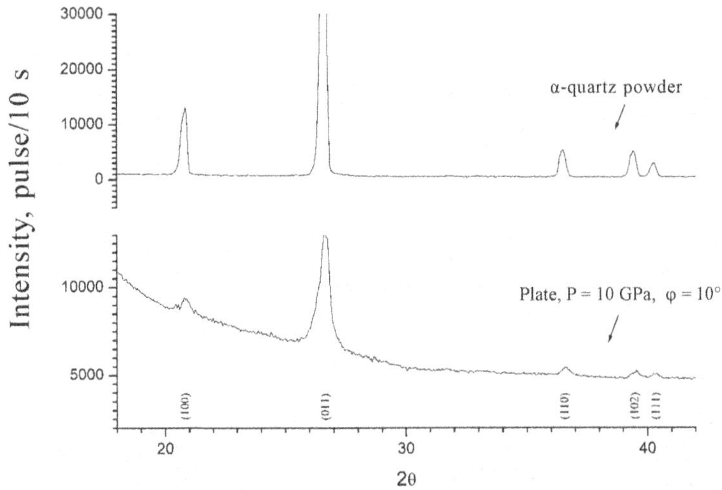

Fig. 75. The diffraction pattern of crystalline quartz: the original powder and the plate after SPD by torsion (pressure 10 GPa, rotated through 10°).

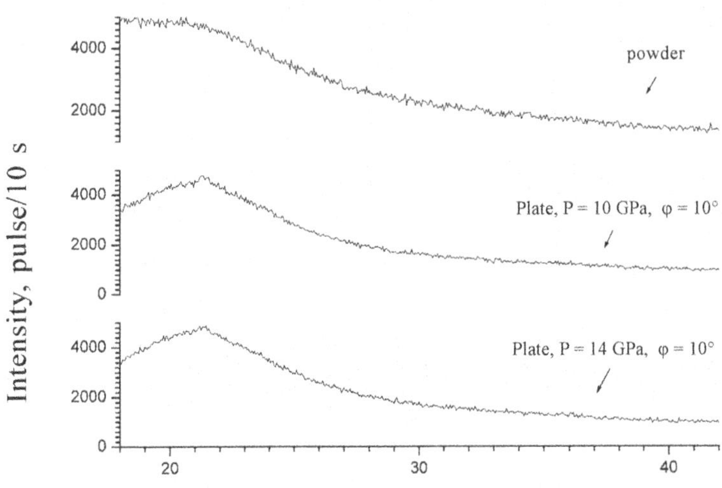

Fig. 76. The diffraction pattern of quartz glass after SPD by torsion (pressure 10 GPa and 14 GPa, rotated through 10°).

diffractogram of a sample of a plate obtained by consolidating the above mentioned α-quartz powder by a rotation of φ = 10° under a pressure of 10 GPa.

We should note a significant (about 1.5 times) line broadening, as well as a higher background level for the consolidated sample. No other phases were recorded for this sample.

As can be seen from Fig. 76 on the fragments of diffractograms, the consolidation of quartz glass powder both at a pressure of 10 GPa and a higher pressure of 14 GPa, does not lead to the appearance of any crystalline phases.

4.3.4. Glasses (slide, quartz)

Glass is an inorganic isotropic substance, a material known and used

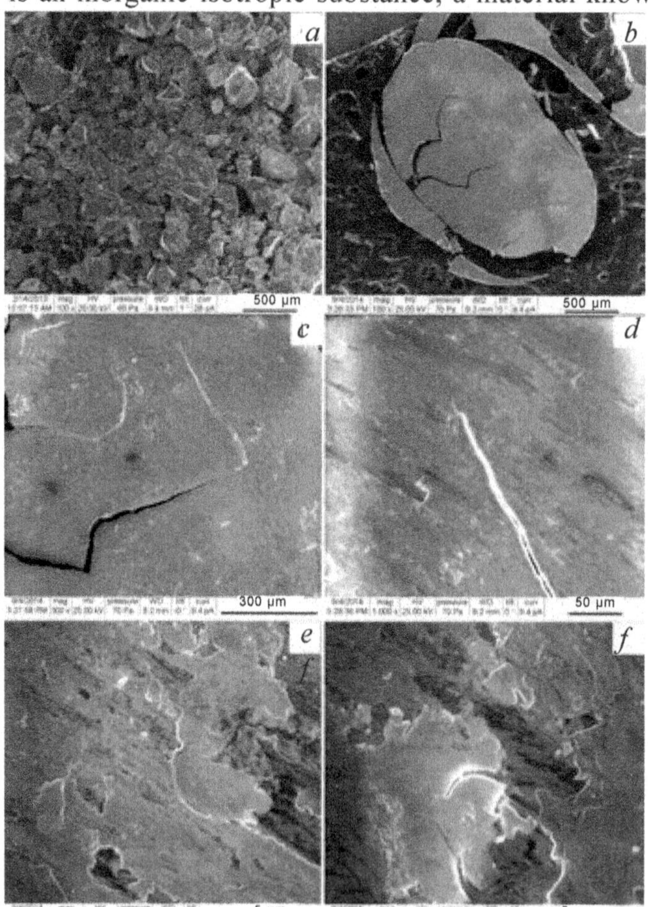

Fig. 77. The surface structure of the slide after SPD by torsion (pressure 10 GPa, rotate through 10°): *a* – the original powder; *b* – plate; *c, d* – macrocracks with different magnification; *e, f* – thin curved microcracks at high magnification.

since ancient times. Glasses are formed by supercooling melts at a rate sufficient to prevent crystallization. Due to this, the glasses usually retain an amorphous state for a long time. It is the structure of glass as an amorphous substance that causes its fragility, which is its main drawback. The structure of amorphous glass is formed by rigid tetrahedra of Si–O bonds with Si atoms in the centre. In the present work, samples are studied after SPD by torsion two types of glass: slide glass (slides) and quartz glass.

Slide. For a slide, two SPD by torsion modes were used, the parameters of which are shown in Table 9. The microstructure of

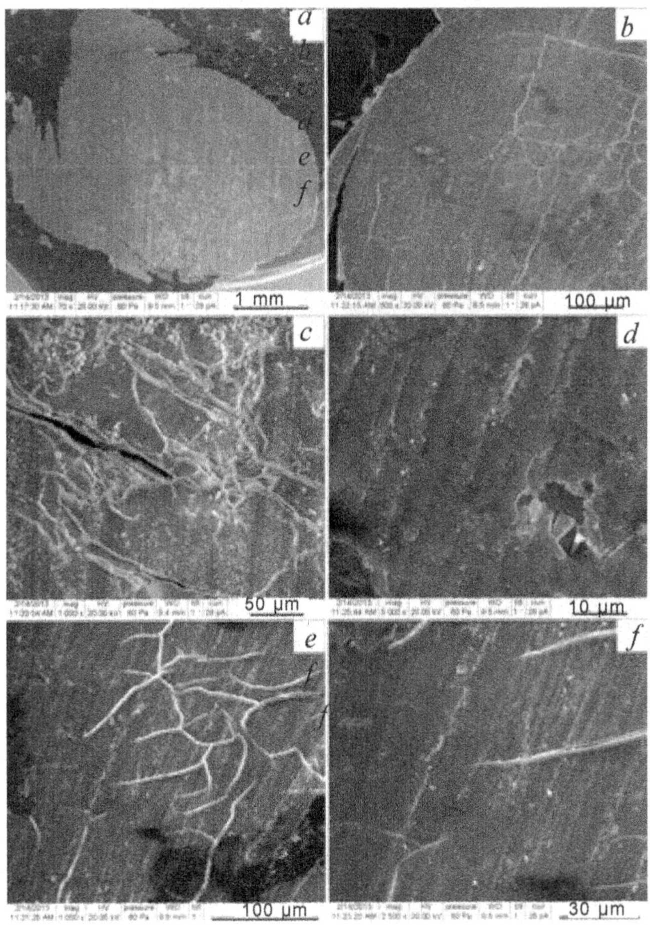

Fig. 78. The surface structure of a slide glass after SPD in torsion (pressure 12 GPa, rotate by 15°): a – plate; *b–d* – thin, curved microcracks with different magnification; *e, f* – the microcracks consisting of rectilinear segments.

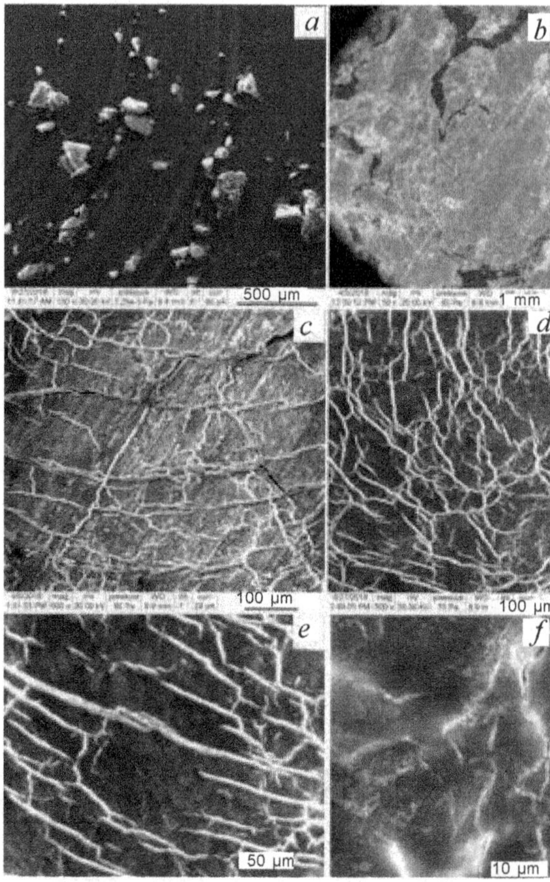

Fig. 79. The structure of the surface of quartz glass after SPD by torsion (pressure 14 GPa, rotated 10°): *a* – the original powder; *b* – plate; *c, d* – branching of microcracks, triple joints; *e* – microcracks consisting of straight segments; *f* – particle formation and crack filling.

the slide in the first case is shown in Fig. 77, in the second – in Fig. 78. SEM images of the powder are shown in Fig. 77 *a*. SPD by torsion in the first mode produced a plate (Fig. 77 *b*), from the edge of which macrocracks propagate. With a larger magnification, it is clearly visible how a microcrack is emitted from the top of the macrocrack (Fig. 77 *c*), the structure of which is visible in Fig. 77. Fine microcracks at high magnification are visible in Fig. 77, *e, f*.

When using the second mode, the plate was also obtained (Fig. 78 *a*). Although the plate is quite perfect and does not contain macrocracks, it does contain microcracks, which are clearly visible at different magnifications in Fig. 78, *b–d*. The picture presented in

Fig. 78 d: microcracks, consisting of straight segments. The structure of a microcrack is visible in Fig. 78 *f*. It can be assumed that the similarity of the images of microcracks in Fig. 78 *b* with those observed in crystalline quartz (Fig. 73 *d*), most likely, is evidence of certain structural transformations in amorphous glass with the second more intensive mode of SPD by torsion. Note that this transformation occurs only in certain regions of the amorphous phase.

The images of the surface of the plate obtained above for different materials show that there is an unambiguous connection between the shape of microcracks and the structural state of the material. The key figures in this regard are Fig. 73 *c* and 78 *b*. These are microcracks consisting of rectilinear segments for crystalline quartz (Fig. 73 *c*) and thin curved microcracks for an amorphous glass slide (Fig. 78 *b*). Further, these key figures will be used in determining the structural state of quartz glass.

Quartz glass. It is considered established that the X-ray diffraction patterns of quartz glass are best interpreted within the framework of the model of a continuous random network of SiO_4 tetrahedra. At the same time, silicon atoms surrounded by four oxygen atoms characterize a short-range order in the glass structure. For comparison in Fig. 72 *a, b* schematically shows the structure of crystalline quartz and the structure of glassy-like quartz in the form of a random network.

Quartz glass can be obtained by melting natural species of rock crystal, veined quartz and quartz sand, as well as synthetic quartz. This substance has other names, for example, fused (amorphous) quartz. Quartz glass has high heat resistance, thermal resistance, dielectric properties and therefore has a very wide scope: optical devices, lamps (gas-discharge, halogen, capsular, etc.), quartz tubes, etc. Extremely pure quartz glass is used for the manufacture of optical fibres in the creation of the fiber optic communication lines.

SPD by torsion of quartz glass was carried out at a pressure of 14 GPa and rotated by 10°. The results are presented in Fig. 79.

Note the similarity of the SEM images shown in Fig. 79, with those obtained above for crystalline ceramics: quartz (Fig. 73) and rock crystal (Fig. 74). In all these cases, there is a branching of microcracks, triple joints, rectilinear segments. The remarkable image of the triple junction in Fig. 79 *d* for quartz glass seems even more convincing than similar images for quartz (Fig. 73 *c*) and the rock crystal (Fig. 74 *d*). The mentioned similarity is all the more surprising since, according to diffractograms (Fig. 76), quartz glass is amorphous in the initial state and remains amorphous after torsion under pressure.

Thus, a paradoxical situation arose. On the one hand, after SPD by torsion, quartz glass is X-ray amorphous, on the other hand, the observed microcracks have features inherent in the crystalline state. An experiment was proposed to clarify a contradiction, which, in our opinion, is apparent.

It has been suggested that the surface layer of a plate of initially amorphous quartz glass undergoes crystallization under SPD by torsion. This is facilitated by the fact that the surface layer is pressed against the anvil. It can be assumed that, when removed from the surface, the plate retains the amorphous state of the initial powder. And then, if in any way we dissolve the surface layer of quartz glass, then on the new surface layer we can observe networks of microcracks inherent in the amorphous state. This is a rather complicated experiment. Nevertheless, we managed to etch the outer

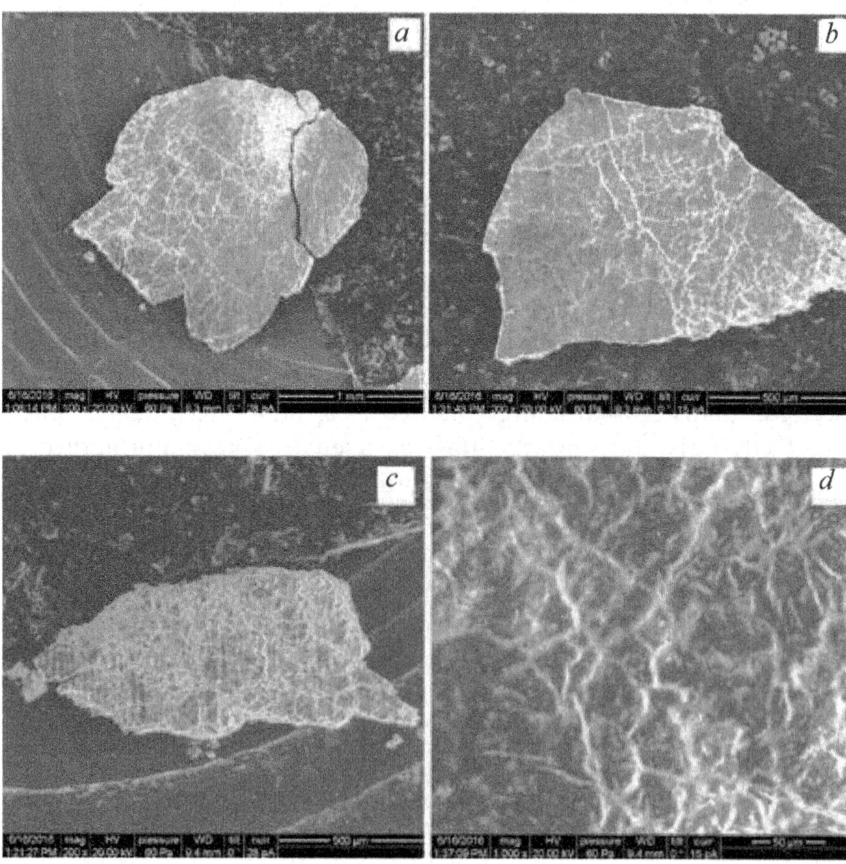

Fig. 80. The structure of the surface of quartz glass after SPD by torsion (pressure 14 GPa, rotated by 10°) and subsequent etching: *a* ... *d* – with different magnification; *e* – a schematic representation of the initial plane (blue colour) and a curved surface (red color) after turning; *f* – glass painting.

layers of quartz glass. Hydrofluoric acid was used. The etching rate is about 100 μm/hour (in a strong solution of hydrofluoric acid).

4.3.5. Glass sticking

Fig. 80 shows SEM images of microcracks on the surface of the plate remaining after etching (Fig. 80 *a*).

Already in this figure and in the subsequent ones it can be seen that microcracks differ significantly from those observed above for the outer crystalline layer of quartz glass (Fig. 79). For microcracks in Fig. 80 there is no characteristic division into rectilinear segments. At the same time, the SEM images in Fig. 80 are strikingly similar to those observed for a slide glass (Fig. 78). We draw attention to the similarity of Fig. 80 *a, b* and 78 *a, b*. In both cases, these are thin, curved microcracks characteristic of amorphous material. Areas with a high density of microcracks in Fig. 80 *a* and *c* appear white. One such area is shown in Fig. 80 *d*. Within this region, the distribution of microcracks is also non-uniform and contains areas almost without microcracks. Thus, it was confirmed that the outer layers crystallize, while the inner layer remains amorphous.

The propagation of cracks for the processes considered in this work occurs at high pressure, preventing the crack from opening. The plate that arises in this case most likely has a gradient structure due to the fact that one plate surface is pressed against the lower rotating anvil, and the other against the upper, stationary. Although the initial material is brittle, but with SPD by torsion a partial consolidation of the powder occurs, free volume remains, due to which the material acquires some viscosity and then becomes brittle again with a large number of revolutions. As noted in [55], uncorrelated accumulation of microcracks and their coalescence lead to the formation of a main macrocrack, which has a single non-branching surface capable of avalanche propagation.

In addition to the gradient structure, there is another consequence of the fact that one of the anvils is fixed and the other is movable. Before torsion, if we select some cutting plane inside the sample and divide it into parts, then all of them will have the same normal. But during torsion, different parts experience different turns depending on their location. This means that the normals to them will have different directions. The curved surface that arises here is shown schematically in Fig. 80 *e*. It is highlighted in red, and the original plane is in blue.

In [57], attention was first paid to the fact that with SPD by torsion the crystallographic plane turns into a curved surface. When studying the evolution of the dislocation structure of the Ni_3Ge intermetallic compound under pressure torsion, it was found for the first time that torsion makes automatic blocking impossible with subsequent heating without a load. We call self-blocking the transformation of dislocations in the absence of external voltage from the sliding configurations to the blocked ones. As applied to the Ni_3Ge intermetallic compound, the auto-blocking is due to the departure of the slip leading to the plane even at zero external stress. It is the transformation of the cross-slip plane into a curved surface that causes the dislocation self-blocking to disappear during SPD by torsion. It is surprising that, despite numerous works on SPD by torsion, this transformation of the crystallographic plane into a curved surface turned out to be unnoticed.

So far we do not know what the consequences can be (when applied to crystalline ceramics) of the transformation of a plane into a curved surface. For example, this may affect the internal structure of microcracks, their splitting into straight-line segments, collisions of microcracks with each other, their distribution, etc.

It was found that SPD by torsion accelerates the consolidation of ceramic powder (quartz, rock crystal, amorphous glass). This is due to the acceleration of such processes as the transfer of particles, their rotation, mutual collisions, interaction with the environment. These processes can be called 'crowd syndrome'. In addition, we believe that the effect known as 'glass sticking' is essential for the consolidation of the powders. In this case, it is valid for each of the studied materials, not only for glasses. First a few words about the effect. It is about sticking together two smooth glasses that are difficult to tear from each other. It is believed that sticking is the result of the action of van der Waals forces, which are rather weak and short-range. Therefore, a large contact area contributes to sticking together. As soon as the adjacent particles with parallel faces, it becomes possible for them to stick together. Adhesion of particles is promoted by their grinding with an increase in the number of revolutions during SPD by torsion.

Above, we paid attention to the fact that the bending of the plane under SPD by torsion (Fig. 80 *e*) was unnoticed, despite numerous works in this area. Similarly, when consolidating ceramic powder (glass, rock crystal), it is natural to expect that the adhesion of glasses will play a dominant role. However, this effect was not involved at all earlier for an explanation of consolidation.

'Sticking of glasses' was already known in the past. In China, artists created glass paintings: the drawing was applied to one surface of a plate, which was then closed with a second plate. Glass paintings were perfectly preserved.

Figure 80 *f* represented a copy of the painting 'Fishermen', which was written by the artist U. Zhen in 1345. The copy is kept in the Art Museum of Shanghai. The painting was created a few decades before the icons of Andrei Rublev.

4.3.6. Microcracks

We believe that the transformation of a powder into a solid material under SPD by torsion occurs as a result of the GRF process, which includes the crushing of particles and their consolidation. In all the cases considered above, the plate that appears after the SPD consists of microvolumes separated by microcracks. In this case, no grains are observed, apparently due to the absence of dislocations. In contrast to the GRF, when welding by explosion, in the case of SPD by torsion, particle scattering and melting are also not observed.

Thus, if as a result of some strong impact there is crushing of particles (powder or solid material), then in order to discontinue to obtain solid material, it is necessary to consolidate the particles arising in the crushing process. At the same time, it is obvious that the PDD processes for various strong effects are not identical. It is therefore of interest to compare further the structure, for example, of quartz after SPD by torsion with the structure of quartz after explosive welding by a metal-metal containing quartz particles.

The transformation of the powder into a solid is possible if excess elastic energy is removed from the indicated region by means of any mechanisms. At the same time, dissipative structures arise, and such that from all possible directions of development choose the one for which the free energy release rate, and first of all the rate of formation of new borders between fragments, will be maximum.

First of all, such a reset mechanism may be the formation of new boundaries between fragments. In fact, most likely, a sequence of dissipative structures arises, to which the elastic energy of explosive action mentioned above is expended. The GRF (granulating fragmentation) is a powerful channel for dissipating the energy input, due to the large total area that the surface of the particles has when they are crushed. Another important dissipative channel is the formation of microcracks, in which the free energy is released due to the growth of the crack surface.

In all the cases discussed above, the initial powder particles have faces, so that when the particles with parallel faces collide, the adhesion of the particles mentioned above (with different orientations) can occur if their surface is smooth. Otherwise there will be a crack. The competition of these processes, namely the adhesion of particles and the formation of cracks, determine the possibility of plate formation.

As can be seen from the above-mentioned drawings, the material is literally penetrated by microcracks. Section 4.1 has already discussed the question of the origin of microcracks and the reasons hindering their transformation into macrocracks. Griffiths [45] formulated the criterion for the start of an already existing crack, based on the balance between the work absorbed by crack surface formation and the elastic energy of the medium released as it grows. Therefore, under constant load the spread of sufficiently small cracks is impossible, with large sizes it is inevitable. In other words, cracks having a length greater than critical grow spontaneously. However, since the torsion is performed under compression, the observed cracks are not, strictly speaking, Griffiths cracks, despite the fact that ceramics and glass are very fragile materials. With a strong compressive effect on such materials, accompanied by intense shear stresses, shear cracks of two types can occur: transverse and longitudinal shear. However, the specified criterion is valid, albeit in a slightly different form than for the Griffith cracks, and with different critical stresses. It should be noted that tensile stresses may still occur during SPD by torsion, but during unloading, so that Griffiths cracks may be among the numerous microcracks.

The features of the network of microcracks are explained in the framework of the approach proposed in [54, 55], in which branching is considered as a manifestation of the dynamic instability of a crack with an excess of incoming energy. The observation of long microcracks is not surprising, given their spontaneous growth under the conditions stated above. The existence of a configuration of microcracks of the triple joint type is also the result of the excess of the released energy over the absorbed. Such an excess may be enough so that instead of one crack it would be enough for two. Characteristic configurations, such as zigzags, result from the rotation of a crack, while others occur when a growing crack stops as a result of a collision with a crack that has arisen earlier. The intricate structure of the grid may also be due to the fact that the corresponding free surfaces experience braking when they meet each other.

4.3.7. Conclusion

The main issue in analyzing the transformation of a powder into a solid material is the following: why the powder particles stick together. For metal powders, the natural mechanism is the joint plastic deformation of the contacting particles, accompanied by the restructuring of the dislocation structure.

In order to clarify the possibility of consolidation in the absence dislocations were carried out electron microscopic studies of a number of fragile materials subjected to SPD by torsion: crystalline materials (quartz and crystal), amorphous materials (glass, object and quartz). Consolidation of powders was discovered, resulting in plates of solid material containing numerous microcracks. In the absence of dislocations, it is just such a network of microcracks filling the entire plate that is a stress relaxation channel, and a significant discharge of the supplied energy falls on the free surfaces belonging to microcracks.

In all the above cases, the particles of the original powder have faces, so that when particles collide with parallel faces, particles can stick together if their surface is smooth. Otherwise microcracks will form.

In quartz, typical features of the branching of microcracks are clearly visible: with a sufficiently large magnification one can see that the microcrack consists of straight segments. It can be assumed that they are located in the cleaved planes. It is of interest to observe in certain places of the crumb that occurs during crushing, and partial filling of microcracks with it. For a plate of crystal obtained by consolidating the powder, the same features of microcrack branching were found. Attention is drawn to the fact that with SPD by torsion the crystallographic plane turns into a curved surface (Fig. 80 *e*). This transformation may affect the internal structure of microcracks, their splitting into straight-line segments, collisions of microcracks with each other, their propagation, etc.

The key images in this regard are the images of a microcrack, consisting of the rectilinear segments mentioned above, for crystalline quartz and the image of thin curved microcracks for an amorphous glass slide.

However, a paradoxical situation arose in the study of quartz glass. The microcracks on the surface of the initial plate, which arose during the consolidation of the powder, have the shape typical of crystalline ceramics, while the sample is amorphous in accordance with the diffractograms. The outer layers were removed by dissolving in hydrofluoric acid. It turned out that microcracks inherent in amorphous material are observed on the surface of the remaining plate. Thus, the original quartz glass plate is a sandwich consisting of the outer layers of crystalline quartz and the inner layer of amorphous quartz glass.

The main results obtained in Section 4.3 can be summarized as follows:

1. A consolidation of powders during torsion under pressure of ceramics and glasses was detected.

2. The association of powder consolidation with granulating fragmentation (GRF) has been revealed.

3. The role of surfaces arising during fragmentation and in the formation of microcracks has been revealed.

4. Dissipative channels that compete with fracture were proposed.

4.4. Surface relief: cusps

The cusps on the interface surface were observed in all the compounds investigated by us, although they had a different shape. For ($\mathbf{C}_p$) copper–tantalum and ($\mathbf{E}_p$) aluminium–tantalum welded joints, having a flat interface, Fig. 30 shows the SEM images of the longitudinal sections of the transition zone. Their similarity is obvious, although the vapours of the starting metals have different mutual solubilities. Here randomly distributed spots are observed, which, as shown by the results of chemical analysis carried out using SEM, correspond to three types of areas: tantalum, copper (aluminium) and zones containing both elements. The tricolour nature of the SEM images can be considered direct evidence of the cusp.

The geometry of the cusps indicates that they are formed by a metal possessing the greatest hardness in this pair. In this case, it is tantalum, in other compounds – aluminide, titanium, iron. As shown by the SEM analysis of the chemical composition, the cusps practically do not contain the second metal, regardless of whether mutual solubility exists. In fact, in the case of cusps, this is not mixing, but interpenetration of materials. The lack of mixing inside the cusps is especially clearly seen for the cusps of tantalum on the flat interface, which correspond to white spots on the image of a longitudinal section (Fig. 30), while mixing leads to a change in the colour of the spots in the local melting zones. The lack of mixing inside the cusps due to the fact that due to the fact that the diffusion in the solid phase is difficult.

Alignment of the cusps was observed for both welded joints ($\mathbf{C}_p$) and ($\mathbf{E}_p$). Figure 30 shows that the cusps are oriented along some selected direction. Especially clearly the tendency to align the cusps in rows is visible in Fig. 29. It can be assumed that the transition to a wave-like surface observed during the intensification of the welding regime occurs as a result of the self-organization of many cusps. The self-organization processes are discussed in detail in Chapter 11.

On the wavy surface, in turn, the cusps are also observed. For the ($\mathbf{C}_w$) welded joints Fig. 36 shows a panorama (optical micrograph), convincingly illustrating the diversity of the cusps. The SEM image of the cusps on the cusps is shown in Fig. 38. Considering the shape of the cusps observed in the longitudinal section, it can be assumed that the cusps have the shape of a cone or pyramid with either a sharp or a smooth top, and they are perpendicular to the surface. Schematically, the cusps are shown in Fig. 81

However, it turned out that on a flat interface the cusps have a different shape. In order to find out which relief the tantalum surface has for the ($\mathbf{C}_p$) welded joint, the following procedure was used.

The copper was etched, and then SEM images of the tantalum surface were obtained at different angles with respect to the beam.

For explanation Fig. 82 *a* shows the surface of a cone with an axis parallel to the beam. For a plane tangential to the conical surface shown in Fig. 82 *a*, shows characteristic angles – the angle of inclination φ and the angle of rotation ω are shown. Some of the SEM images are shown in Fig. 82 *b–d*. In fact, each of the Fig. 82 *b–d* gives a view of the surface of tantalum from the point of view of various observers. In other words, SEM micrographs represent different perspectives, at which a combination of three-dimensional objects, such as cusps, is observed.

We draw attention to the essential feature of the image of the cusps: the repeatability of self-similar elements. One of these elements is shown in Fig. 83. In fact, this is an SEM image of several cusps. It is the self-similar nature of the cusps that was one of the factors that initiated the fractal approach to the description of the structure of the interface, presented in Chapter 6.

Although the interface is flat, the cusps are similar to splashes of waves. Such a similarity (see Fig. 82 *e, f*) is unexpected, considering that the cusps are formed by a solid phas
e. Moreover, this phase did not experience melting. The observation of the splashes is to a certain extent a confirmation of the assumption made above that the cusps serve as precursors of the wavy relief.

A similar topography of the interface was observed for the aluminium–tantalum welded joint ($\mathbf{E}_p$). Figure 84 shows the SEM images of the tantalum surface at different magnifications (aluminium was etched).

The same picture was observed for the magnesium–titanium welded joint. We investigated the structure of this welded joint in order to ascertain the possibility of magnesium boiling during explosive welding, leading to the formation of pores (see Fig. 88).

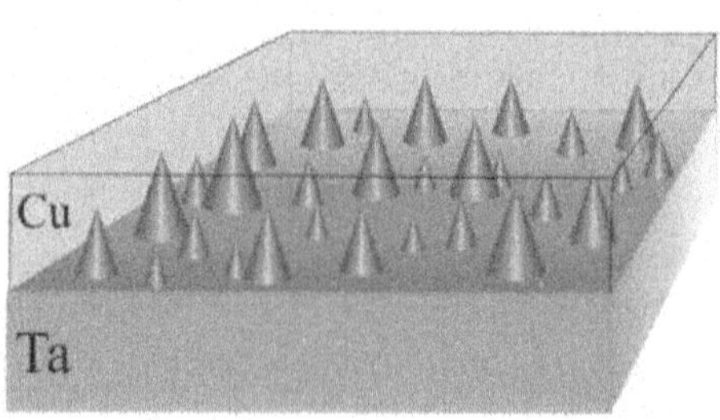

Fig. 81. Schematic representation of cusps.

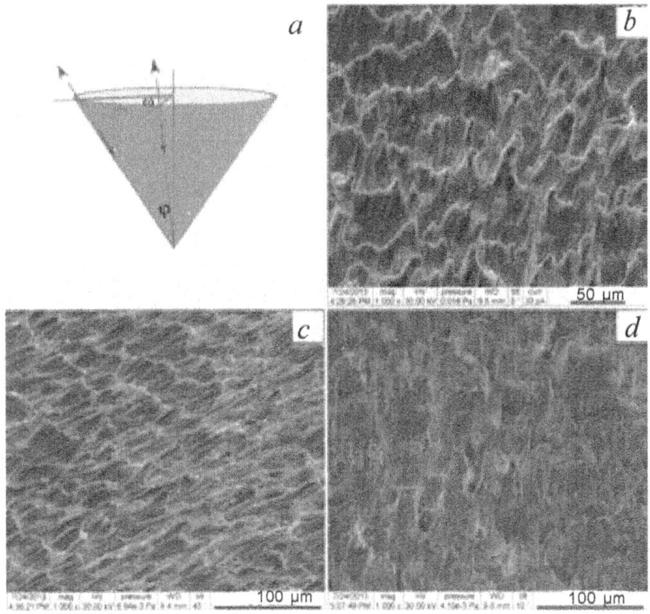

Fig. 82. SEM image of the tantalum flat surface for the ($\mathbf{C}_p$) welded joint (copper etched) at different values of the angle of inclination φ and the angle of rotation ω: *a* – scheme; *b* - without tilting; *c* – φ = 45°, ω = 45°; *d* – φ = 0, ω = 180°; *e* – φ = 45°, ω = 90°; *f* – splashes of waves (from B. Akunin's blog 'Love for History').

For the Mg–Ti welded joint, which has a flat boundary, no pores were observed. In this case, the surface of titanium, after magnesium was etched, contained splashes.

The SEM images of the titanium surface at different angles are shown in Fig. 85. Here, as before, the alignment of the splashes along the selected directions is clearly visible.

Thus, the cusps are observed, regardless of what is the mutual solubility of the starting metals, and on what form (flat or wavy) the interface has. If the interface were smooth, there would be problems with adhesion, and either a reconstruction of the metal bonds or the transport of point defects would be required. But the presence of cusps contributes to the coupling: the cusps play the role of 'wedges', connecting the contacting materials with each other.

4.5. Melting

According to Deribas [1], the diversity of the structures of the transition zone in different welds includes: areas where there is

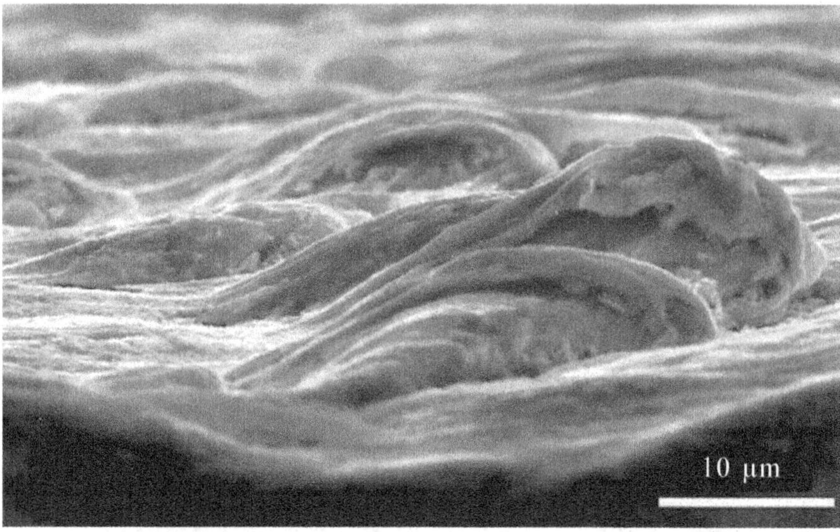

Fig. 83. SEM image of the tantalum flat surface for the (C_p) welded joint. (copper etched out).

no melt, hardened melts, small interlayers of melts at the interface (titanium–steel), a large number of particles of one metal trapped in another or melt (copper–niobium), areas where there is no visible mixing zone but there is a smooth change in the concentration of all elements in a zone several microns wide (low-carbon steel–stainless steel), areas with intermittent components and narrow interlayers of the average composition there between (nickel–steel). Judging

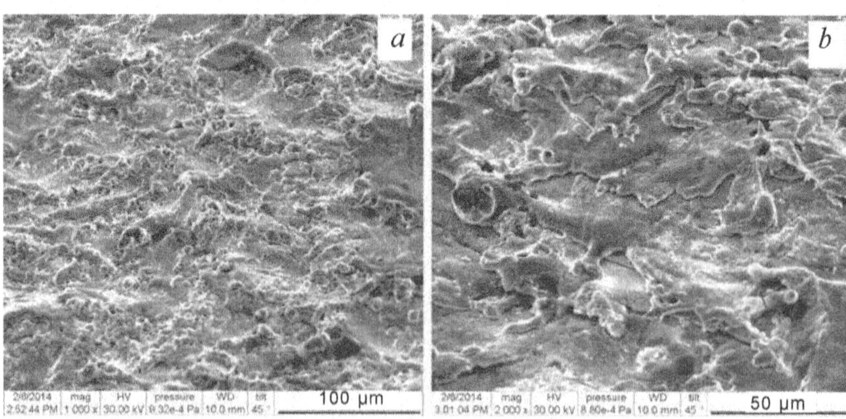

Fig. 84. SEM image of the tantalum flat surface for the (E_p) welded joint (aluminium etched): *a, b* – with different magnification.

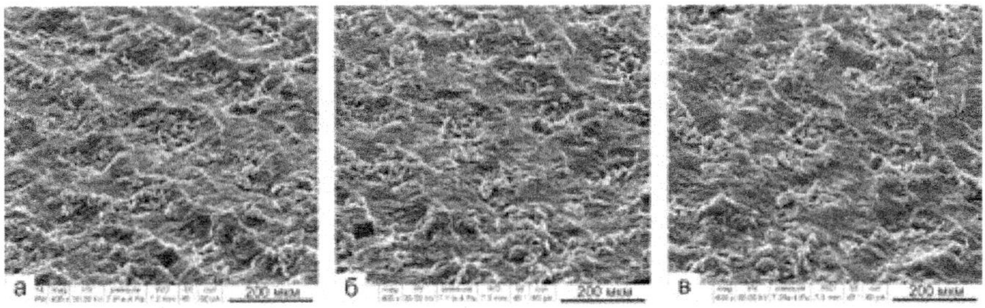

Fig. 85. SEM images of the surface of titanium for the magnesium–titanium welded joint (magnesium etched) at different values of the angle of inclination φ and angle of rotation ω: *a* – φ = 45°, φ = 30°; *b* – φ = 45°, ω = 60°; *c* – φ = 45°, ω = 90°.

by the diversity of the structure of the transition zone in different compounds, there are different scenarios for their formation.

There are different points of view on the phenomena occurring during explosive welding. Fragmentation, similar to that observed during severe deformation, was considered the only type of fragmentation. Granulating fragmentation (GRF) did not seem so obvious, but the similarity with the Mott fragmentation during the explosion and the observation of the scattering of particles such as fragments turned out to be quite convincing arguments in favour of the GRF. The notion of surface heterogeneity as an alternative to an atomically smooth surface is gradually becoming generally accepted. Detailing of heterogeneity was not carried out. We have identified a significant role in the adhesion of materials such surface inhomogeneities, which we called the cusps. But melting during explosive welding always caused a negative attitude and the desire to overcome it. We cannot agree with this, considering the results of electron microscopic observations.

4.5.1. Particle scattering and melting

Melting was observed in various forms: either in the form of a film along the entire interface, or in the form of separate parts of the film, or in the form of zones of local melting. The film was observed for welded joints ($\mathbf{A}_p$) and ($\mathbf{E}_w$), the partially melted boundary has the welded joint ($\mathbf{B}_p$), local melting zones were observed in all other welded joints. Obviously, we are talking about molten, but then frozen forms. We confine ourselves to listing only typical micrographs. The internal structure of the film for the welded joint ($\mathbf{A}_p$) is shown in Fig. 23, and the partially melted boundary for the welded joint (**B**$_p$) is shown in Fig. 26. As noted above in section 3.1, the spherulites in Fig. 23 and dendrites in Fig. 26 are witnesses of melting for both welded joints (**A**$_p$) and (**B**$_p$), respectively. Spherulites were also observed for the welded joint (**E**$_w$) (Figs. 46, 47).

The microheterogeneous structure of the local melting zones is shown in Fig. 33 for the welded joint (**C**$_p$) and in Fig. 38 *b* for the welded joint (**C**$_w$) of metals that do not have mutual solubility. Regardless of whether the interface is flat or wavy, the local melting zone is filled with a colloidal solution (suspension type) of tantalum in copper, in which the dispersion medium is molten (and then frozen) copper, and the dispersed phase is tantalum, which does not undergo melting. The difference between the structures of the zones is only in the sizes of tantalum particles, which was discussed above.

GRF is an analogue of fragmentation in an explosion, but occurring in the presence of various strong barriers. For particles (fragments), emitted during explosive welding from one plate, both the second plate and the remaining bulk of the original plate will serve as such barriers. In this case, it can be assumed that the expansion of solid particles of a phase that does not undergo melting will initiate a local melting of a more fusible material near the interface. This is due to the fact that, due to the large total area of the particles, the effective friction between the particles and the barrier can cause local heating, sufficient for melting. To a certain extent, such scattering particles are analogues of sparks, which can often be observed under conditions of strong dry friction between materials.

In explosive welding, this mechanism works regardless of whether the initial phases have mutual solubility. For immiscible phases, the result of its action may be the microheterogeneous structure of local melting zones (Fig. 33, 38 *b*).

The following scenario is possible: explosion, particle formation, local scattering of particles (fragments), interaction of particles with an obstacle, local melting, preservation of a fragmented layer of non-interlaced particles. The shape of the particles inside the local melting zone is significantly different from the shape of the particles inside the fragmented layer. For the (**A**$_w$) welded joint, the aluminide particles inside the local melting zone (Fig. 20 *b*) are fairly smooth, while within the fragmented layer (Fig. 18 *c*) are more rugged. Similarly, in the (**C**$_w$) welded joint, scattered tantalum particles (Fig. 38 *c*) are observed to be significantly less rugged than those that did not fly (Fig. 69 *c*). The question remains as to why particles like fragments are sufficiently smooth. It is natural to assume that the scattering of smooth particles is an easier process than the rugged ones. Therefore, smooth (or rather less rugged) particles scatter, while the rugged ones remain on the edge of the refractory phase. In the general case, the convex shape of particles contributes to their expansion, whereas the presence of concave segments in the outline of rugged particles, on the contrary, keeps them from expanding.

The connection between the scattering of aluminide particles and the local melting of titanium for the ($\mathbf{A}_w$) welded joint is illustrated in Fig. 20 *a*. In this case, numerous aluminide particles are surrounded by a solid solution of titanium doped with niobium and aluminium, which are originally part of the aluminide.

For welded joints of aluminium and tantalum, the structure of the local melting zones looks somewhat different. As shown above, for the ($\mathbf{E}_w$) welded joint (Fig. 46, 47), as a result of the interatomic interaction of aluminium and tantalum, intermetallic reactions become possible, leading to the formation of cubic Al_3Ta particles. However, for the ($\mathbf{E}_p$) welded joint the diffraction pattern (Fig. 43 *a*) does not show the presence of intermetallic compounds. In Fig. 86 the ($\mathbf{E}_p$) welded joint shows clearly visible particles without faceting, with sizes of 50...150 nm, randomly located in the molten layer. Particles in Fig. 86 appear to be clusters.

4.5.2. Colloidal solutions

In cases where the starting metals do not have mutual solubility, the local melting zones are colloidal solutions in which the particles of the starting elements are mixed. Consider the conditions under which certain forms of colloidal solutions occur [58]. We confine ourselves initially to the case when during explosive welding no one of the materials boils and the gas phase does not form, respectively. The relationship between the following temperatures is significant:

T_m is the melting point of the metal (1),

T_m is the melting point of the metal (2),

T_s is the temperature near the interface.

The various options are shown schematically in Fig. 87 assuming (for simplicity) that $T^{(1)}_m > T^{(2)}_m$. Consider the case (Fig. 87 *a*), when

$$T^{(1)}_m > T_s > T^{(2)}_m \qquad (2)$$

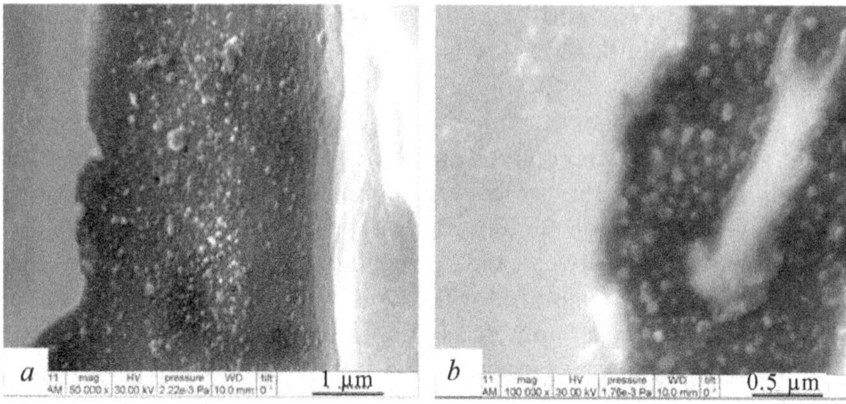

Fig. 86. SEM image of tantalum surface for ($\mathbf{E}_p$) welded joint: *a, b* – with different magnification.

When performing (2), a melt of the low-melting phase appears (hatching in Fig. 87, a), containing solid particles of the refractory phase. This is a colloidal solution of the suspension type. Due to the circulation, the colloidal solution is mixed, which upon subsequent solidification becomes a frozen suspension, dispersion strengthened by particles of the refractory phase. In this case, the local melting zones not only do not pose a danger to the welded joint, but, on the contrary, can contribute to its hardening.

Consider the case (Fig. 87 *b*), when

$$T_s > T_m^{(1)} > T_m^{(2)} \tag{3}$$

This means that in the interval (1) $T_s > T_m > T_m^{(1)}$ both metals are melted (double hatching in Fig. 87 *b*). Due to the strong external influence both liquids break up into drops, a colloidal solution of the emulsion type appears. But the emulsion, generally speaking, is unstable, drops of one kind stick together, so that separation occurs into two immiscible liquids.

In the alternative case (Fig. 87 *c*), when

$$T_m^{(1)} > T_m^{(2)} > T_s \tag{4}$$

both metals do not melt and colloidal solutions are not formed.

Thus, for colloidal solutions of immiscible liquids, either the emulsion variant or the suspension variant is realized. When hardening, the emulsion poses a danger to the continuity of the joint

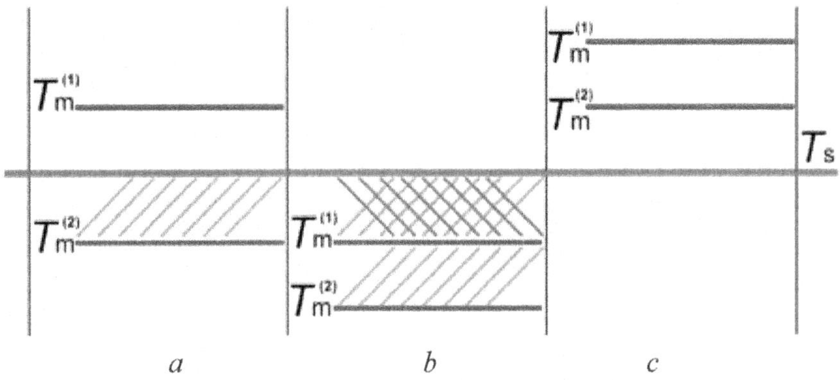

Fig. 87. Schematic representation of various sequences of characteristic temperatures.

due to possible stratification, while the suspension, on the contrary,)may contribute to the strengthening of the joint.

For the copper–tantalum and iron–silver welded joints, the characteristic temperatures are approximately equal to: $T_m^{(Ta)} = 3300$ K, $T_m^{(Cu)} = 1300$ K; $T_m^{(Fe)} = 1800$ K, $T_m^{(Ag)} = 1200$ K. Suppose that the temperature near the interface is $T_s = 1500...1700$ K. Then in both cases relation (2) is formed and a dispersion-strengthened suspension is formed, which provides strength of the welded joints. If $T_s = 2000$ K, then the relation (2) is still satisfied for the copper–tantalum ($\mathbf{C}_p$) and ($\mathbf{C}_w$) welded joints and a suspension is formed, whereas for iron–silver, relation (3) is fulfilled and an emulsion is formed [40].

For the iron–silver welded joint ($\mathbf{D}_w$) the local melting zone consists of silver containing iron particles (Figs. 52, 53). It is significant that the structure of the local melting zone is inhomogeneous: there is a region where the iron particles are almost invisible. This means that there are both concentrated and non-concentrated solutions of iron in silver. However, no delamination was observed for the iron–silver welded joint ($\mathbf{D}_w$) [40].

In explosive welding, not only emulsions and suspensions can be formed, but also another form of colloidal solutions, namely foam. It can be expected that for immiscible phases at a certain ratio of characteristic temperatures, a foam is formed on the interface in which the dispersed phase is gas bubbles of one of the phases, and the dispersed medium is a colloidal solution in the form of thin films. Suppose that relation (3) holds, so that near the interface both metals are melted. If, moreover, for a low-melting metal having a boiling point of (2) TV, the relation

$$T_s > T^{(2)} \tag{5}$$

is satisfied, foam may form near the interface.

Consider a welded joint of metals magnesium and titanium having limited solubility. The melting points T_m of both metals are: T_m(Ti) = 1933 K, T_m (Mg) = 922 K, and the boiling point of magnesium in T_b(Mg) = 1363 K. Due to the low boiling point of magnesium, the molten magnesium can also contain a gas phase. It is considered established that the formation of any stable foam in a pure liquid is impossible. We believe that in those zones where both metals are melted, foam formation is possible, in which liquid layers are formed by a magnesium–titanium colloidal solution, and the bubbles are filled with magnesium vapouur. For the formation of foam, shaling of the liquid is necessary, which is provided during the explosion. Due to ultrafast cooling, the liquid dispersion medium turns into a solid phase, and the gas dispersed phase is preserved. The result is a solid foam with high stability. But this means that pores will be observed near the interface.

In Fig. 88 one can see two magnesium–titanium welded joints, obtained at different modes of explosive welding:

- (*a*) contact speed of 2000 m/s, impact velocity of 740 m/s, impact angle of 21.5°, energy expended on plastic deformation = 0.56 MJ /m^2;
- (*b*) contact speed of 2300 m/s, impact velocity of 860 m/s, impact angle of 21.5°, energy expended on plastic deformation = 0.7 MJ/m^2.

In case (*a*) the boundary is flat, in case (*b*) it is wavy. As can be seen from Fig. 88, the state of the boundaries is different: in contrast to the dense boundary (*a*), the boundary (*b*) is porous. Above, in Fig. 85, splashes on a flat titanium boundary (magnesium etched) are shown.

Pure magnesium is unsuitable due to its low corrosion resistance and low strength. Magnesium alloys are used as a structural material [59]. Of all structural materials, magnesium alloys are distinguished by the lowest density (4 times less than that of steel), which makes them suitable for structures in which mass is the main indicator. We draw attention to the fact that when developing composites it is necessary to take into account the possibility of boiling of

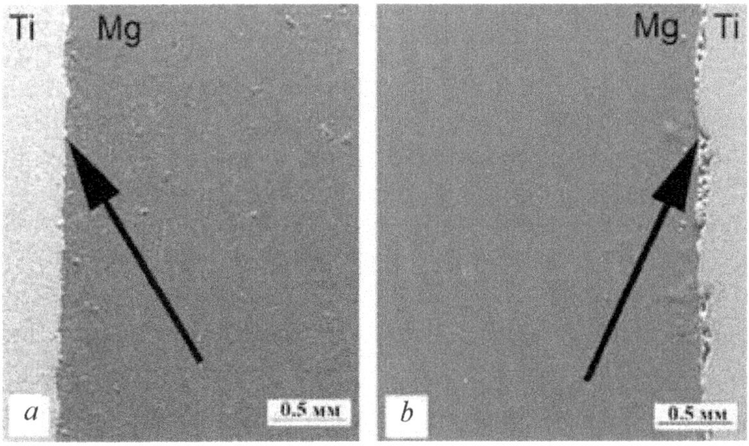

Fig. 88. Magnesium–titanium: *a* – tight boundary; *b* – porous boundary.

magnesium during the process of explosive welding, the formation of foam and the loss of continuity of the welded joint.

4.5.3. Vortex formation

The vortex structure of local melting zones was observed in many welded joints [1, 3]. The problem of vortex formation still does not have a unique solution. Among the welded joints studied by us, the vortex formation is especially pronounced in the local melting zones for the (**A**$_w$) titanium–aluminide welded joints (Figs. 11, 13, 14). The vortices for the (**C**$_w$) welded joints have a less perfect shape (Fig. 37). The results are mainly presented in [23, 24, 60, 61], as well as in other works cited above.

For (**A**$_w$) welded joint in Fig. 11 *a, b*, areas with a vortex structure are clearly visible. Clearly visible are closed layers that form a vortex. Similar areas are visible in Fig. 13. It can be seen that the internal structure of the vortex is rather complicated (ragged layers, small additional vortices). Based on measurements of the composition of the vortex (Fig. 14), data on its chemical composition were obtained, which are listed in Table 10.

In fact, the main part of the vortex zone contains a solid solution based on titanium ($\geq$80% Ti), enriched in niobium and aluminium. TEM analysis showed that the vortex zone is ultrafine (Fig. 16). In fact, the vortex zone is nanocrystalline (Fig. 17). Similarly, an analysis of the structure of the vortices observed in the transition zone of the titanium–VTI-4 ($\mathbf{B}_w$) welded joint was performed. Data on the chemical composition of the vortex zone were obtained using SEM. As for the ($\mathbf{A}_w$) welded joints, the main part of the vortex zone contains a solid solution based on titanium.

For titanium–aluminide welded joints, it is possible to calculate diffusion paths of atoms from one material to another during an explosive time of about 10^{-6} s at a temperature close to 1000°C and compare them with the width of the transition zone. Assuming that the diffusion coefficient D in a solid at a specified temperature is no more than 10^{-1} cm²/s, we find that the free path length of the range of atoms is $\ell \approx 2\sqrt{Dt} \approx 2{\cdot}10^{-1}\,\mathrm{nm}$. This means that due to the extremely short welding time, diffusion in the solid phase should not play any significant role in the formation of the mixing zone. Therefore, the observed above heterogeneity of the interface and the transfer of particles of one material to another can not be explained by diffusion.

Using the typical value of $D \approx 5 \cdot 10^{-5}$ cm²/s for the diffusion coefficient in a liquid, we find that in this case the free path lengths of diffusion mixing are very small: $l \approx 60$ nm. The width of the transition zone during explosive welding, generally speaking, varies from one section to another. As an estimate, one can use the size of the vortex region, which varies within 30...150 μm. These values are several orders of magnitude (at least four) greater than the estimate for the value of l in a liquid. However, it should be noted that the time of existence of the melt is comparable to the cooling time, i.e. significantly more than 10^{-6} s. As a result, the atomic path lengths can be of the order of several microns. However, even such lengths still do not provide the observable dimensions of the vortices. The situation changes if we take into account the dramatic change in the transfer processes caused by turbulent motion, namely, their intensification compared with the corresponding molecular transfer processes, including an increase in energy dissipation, momentum transfer, heat transfer and particle diffusion [62, 63]. Turbulent transfer coefficients for large Reynolds numbers can be $10^5...10^6$ times higher than molecular transfer coefficients. As a result, there is a sharp increase in the mixing of the fluid, which is considered one of the most characteristic features of the turbulent motion.

Based on the above estimates, we can assume that we are witnessing a frozen whirl generated from the melt. This is the macrostructure of the vortex, and the layers are its mesostructure. A mixture of α- and β-grains is a vortex nanostructure. Accordingly,

Table 10. Results of spot measurements of the chemical composition in the vortex zone

Measurement No.	Concentration, at.%		
	Ti	Al	Nb
0	57.00	25.72	17.28
1	82.52	10.71	6.77
2	81.64	11.21	7.15
3	81.94	10.63	7.43
4	81.69	11.04	7.27
5	88.51	07.21	4.27
6	88.51	07.36	4.13
7	98.27	1.27	0.46

there are three characteristic dimensions of the vortex: ~20...100 μm for the macrostructure, ~2 μm for the mesostructure, ~30...100 nm for the vortex nanostructure. From the entire transition region, only in the vortex zone mixing is associated with diffusion, but with turbulent diffusion.

Confirmation that the vortices arise as a result of local melting is the observation of the duplex structure of the vortex (Fig. 16, 17), consisting of clearly faceted α- and β-phase grains of variable composition, and the β-phase was not observed at all in the original materials condition.

Among the numerous vortices known in hydrodynamics, the vortex arising from explosive welding takes a special position: it originates in a closed cavity with moving boundaries without an external source. The source is actually the borders themselves, it is from them that the small vortices are broken, from which larger vortices are organized.

As can be seen from Figs. 13 and 14, the vortices are located near the weld and are clamped between the aluminide and titanium. In most cases, the main part of the vortex zone is in titanium, so that its geometric centre is removed from the interface by about 15...20 μm. Sometimes there is emptiness in the centre. most likely as a result of shrinkage due to the difference in specific volumes of liquid and solid.

The possibility of vortex formation around the 'hot' spot, from where melting begins, in principle, is determined by the convective instability of the liquid. It is known [64] that a necessary condition for the occurrence in the field of gravity of a stable and correct convective vortex in a layer of thickness L and temperature difference at the lower and upper boundaries $\Delta T > 0$ is the Rayleigh number Ra of some critical value $Ra_{cr} \gg 1$. We are limited to estimates for the melt of titanium. Using the values of the thermodynamic parameters of titanium, which are included in the corresponding inequalities, we find that for $L \approx 100$ μm, the Rayleigh number:

$$Ra \leq (3 \div 5) \cdot 10^{-2} \qquad (6)$$

This means that stable vortices cannot arise during the lifetime of the molten state, which is much longer than the blast wave action time. As a result, the observed layered (quasi-vortex) structure of localized zones cannot be due to convection in the melt. The main reason for this is the small value of the characteristic spatial scales – L. In order to bring the value of Ra to Ra_{cr}, it is necessary to increase L by at least two orders of magnitude. But then the size of the melting region should be of the order of mm, which greatly exceeds the experimentally observed thickness of the transition zone of the compound.

Twisting of the layers could be expected around a 'hot spot' that is near the interface. However, as can be seen from Figs. 13 and 14, the layers are twisted not around the 'hot spot', but around the geometric centre of the cavity. An alternative variant of the appearance of a layered (quasi-vortex) structure cannot be ruled out: layer-by-layer propagation of the crystallization of the melt in the radial direction from the boundary of the closed cavity to its geometric centre. This way (for geological times) the formation of a layered structure of agate occurs [65], which is similar to the observed structure of the local melting zone. Indeed, in both cases the rings follow the contour of the cavity. However, they, as a rule, are not observed near its centre, where in some cases emptiness is observed. We draw attention to the recrystallized zone of titanium observed in the vicinity of the interface (highlighted in Fig. 11 *a*, *b*). Its existence means that the duration of the experiment is much longer than the time of the explosion and includes cooling at residual temperatures. But if this time was sufficient for recrystallization of titanium, which occurs in the solid phase, then all the more it is sufficient for the crystallization of the melt. Indeed, due to the above-mentioned difference in the diffusion coefficients in the solid and liquid phases, the time of formation of the recrystallization zone in solid titanium will exceed the melt crystallization time.

Thus, simultaneous observation (near the same wavy border) of the recrystallized zone of titanium and local zones of melting confirm the possibility of layer-by-layer crystallization of the melt.

4.5.4. Melting and bonding

Local melting zones are 'inserts' that have a different structure compared to the environment. Depending on the structure of the 'inserts', they can be or, on the contrary, be not dangerous for the strength of the welded joint. The structure of the local melting zone is not always 'good'. In the case of normal solubility, the formation of brittle intermetallic phases is a hazard. However, it is known that for many materials intermetallic inclusions are used as a hardening phase. Therefore, intermetallic 'inserts' arising from explosive welding are also not always 'bad'. Intermetallic reactions are a real danger if the melt propagates along the entire interface. Then the possibility of obtaining a welded joint depends on the conjugation of the intermetallic phase with both materials. In the absence of mutual solubility of the starting materials of the local melting zone, they are colloidal solutions of immiscible liquids. When hardening, the emulsion poses a danger to the continuity of the compound due to possible delamination, while a suspension containing particles of the refractory phase, in contrast, may contribute to strengthening the joint. In this case, precisely because of the lack of mutual solubility, there is no danger associated with intermetallic compounds, the formation of which is impossible without diffusion. As a result of the research, to a certain extent, the stereotype about the danger of melting during explosive welding was overcome.

The fact that the melted region for the ($\mathbf{E}_w$) welded joint is a film at the interface, whereas for the ($\mathbf{E}_p$) welded joint are isolated zones, is consistent with the intensification of the welding mode used to produce the ($\mathbf{E}_w$) welded joint (see Fig. 1) and evidence of approaching the upper boundary of the 'weldability window'. However, the ($\mathbf{E}_w$) welded joint remains stable, i.e. welding parameters have not yet reached values that are dangerous for the integrity of the connection. Moreover, we believe that in a favourable case (at not too high values of the parameters) melting, on the contrary, promotes the formation of a joint, due to the gluing of surfaces. In this regard, the micrograph (Fig. 46) for the compound ($\mathbf{E}_w$), which demonstrates the transition from aluminium to tantalum through molten aluminium, aluminium with particles, melted tantalum, is extremely convincing. It is obvious that such a transition facilitates the adhesion of aluminium and tantalum compared with a sharp solid-phase transition.

For many materials, in particular for polymers, the best adhesive substance is the solution or melt of this substance [66...68]. This, to some extent, confirms the possibility of bonding the metals under study as well due to their melting during explosive welding. Indeed, during melting, the problems of wetting, adhesion, thermal expansion and protection against contact corrosion are immediately solved.

This is true for adhesion between the molten phase and the solid "one's own" phase. In this case, we are talking about the adhesion between the molten aluminium and solid aluminium, not experiencing melting (black bar on the left in Fig. 46). But this is not obvious for the adhesion of the melt with the 'alien' phase. In this case we are talking about wetting tantalum with molten aluminium. In fact, the question arises of wetting a solid with a liquid for both cases: 'yours is yours' and 'yours is someone else's'. For the aluminium–tantalum ($\mathbf{E}_w$) welded joint, as can be seen from Fig. 46, during the welding process, the tantalum surface is wetted by the melt and the melt spreads over the surface.

In the case of the ($\mathbf{A}_w$) welded joint, as seen in Fig. 8, there are observed local melting zones, which are capsules pressed against the surface of the aluminide. The capsules are located along the entire wavy interface. A similar picture is observed, as can be seen from Fig. 37 *a*, for the copper–tantalum ($\mathbf{C}_w$) welded joint. In this case, we are talking about wetting tantalum with molten copper.

The melt plays the role of a glue. Capsules provide spot bonding, but only if wetting occurs along the entire surface of the capsule, both at the boundary between one's own and one's own and another's. It should be noted that the wetting and spreading processes are facilitated by numerous cusps at the interface, leading to its roughness. In any case, when selecting metals for explosive welding, it is necessary to take into account the degree of wetting of the refractory phase by the melt of the low-melting phase.

It is known that the adhesive film should not exceed a certain thickness. When applied to welded joints, this may mean a limit on the thickness of the molten region. With the intensification of the welding mode and approaching the upper weldability limit, the critical thickness of the molten region will be reached, at which bonding becomes impossible. Thus, explosive welding in one degree or another absorbs another method of joining materials, namely joining them by forming molten areas. As a result of the research, to a certain extent, the stereotype about the danger of melting during explosive welding has been overcome.

The above-mentioned capsules provide spot bonding, but only if along the entire surface of the capsule there is a wetting of both the 'own–own' and 'own–alien' phases. However, the degree of wetting appears to be sufficient along the entire surface of each of the capsules. This is clearly seen in the example of the local melting zones shown in Fig. 37 *a*. Otherwise, it would be impossible to obtain high quality and stability of the chemical reactor wall containing a copper–tantalum welded joint.

5

Risk zones when explosive welding

5.1. Chemical reactor

Figure 89 shows an image of a chemical reactor vessel having a length of 12.6 m, a diameter of 2.1 m, a wall thickness of 50 mm, a layer thickness of tantalum 1 mm [37]. The vessel is made of a composite: carbon steel–copper–tantalum, by means of explosive welding. Works were performed by the Dynamic Materials Corporation (USA). The inner shell consists of a layer of tantalum with a thickness of 1 mm. The whole structure is based on the corrosion resistance of tantalum, as high as its price. And despite this, an enormous structure, covered with tantalum from the inside, was obtained by explosive welding. The outer shell is made of carbon steel. As shown in [37], both internal boundaries of the composite are wavy. For the copper–tantalum interface, the wavelength and amplitude do not remain fixed along the border. On average, they are 320 and 150 μm.

In the case under study, for the copper–tantalum ($\mathbf{C}_w$) welded joints, the corresponding values are approximately ~270...350 μm and ~60...65 μm (Fig. 35). The welding parameters used were not identical. In [37], composite sheets with a size of 1800 mm × 300 mm × 55 mm (see Fig. 89) were obtained. Laboratory samples were used: tantalum (thickness 0.1 mm) and copper (thickness 3.5 mm), the remaining dimensions are tens of mm. However, the reasons for the high quality and stability of the copper–tantalum ($\mathbf{C}_w$) welded joints, formulated below, are quite common. Using the example of laboratory samples, we tried to find out the reasons for the strength and stability of a welded joint with a macroscale.

Fig. 89. The case of a chemical reactor made of a steel–copper–tantalum composite (explosion welding) [37].

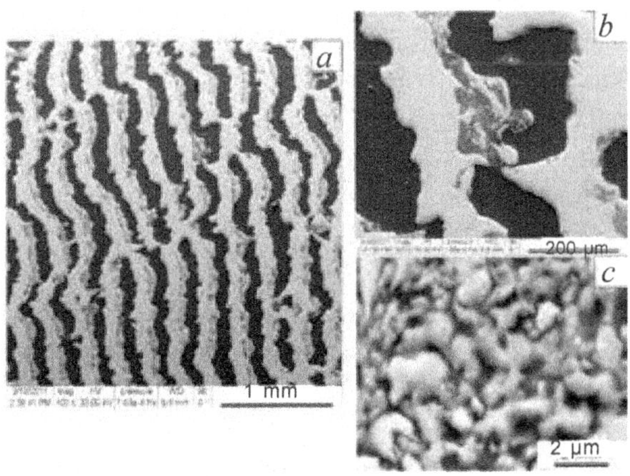

Fig. 90. Longitudinal section of the wavy interface for the ($\mathbf{C}_w$) welded joint: *a* – copper and tantalum bands; *b* – tantalum cusp and local melting zone: *c* – copper–tantalum suspension.

Figure 90 provides sufficient information to determine the reasons for the high quality of the ($\mathbf{C}_w$) welded joint. First of all, Fig. 90 *a* shows a relatively regular wavy interface, containing numerous cusps and zones of local melting. Figure 90 *b* shows a large protrusion of tantalum (about 100 µm) and a local melting zone in the nearest copper band. The role of the cusps fastening the contacting surfaces between them has been discussed above. Due to the lack of mutual solubility of copper and tantalum, the local melting zone cannot be a true solution. For the same reason, there is no diffusion here and intermetallic reactions are impossible. This zone is a colloidal solution. Due to the high melting point of tantalum (3300 K), the colloidal solution cannot be an emulsion and, accordingly, there is no danger of delamination.

These zones are filled with a suspension (Fig. 90 *c*), which consists of a copper matrix containing micron tantalum particles. These zones are the 'capsules' hardened by tantalum particles referred to above, which provide spot bonding of the contacting surfaces (see Section 4.5.4). Thus, the lack of mutual solubility prevents the formation of intermetallic compounds, and the presence of surface irregularities, such as cusps and zones of molten (and then frozen) colloidal solutions, provides a bond between the contacting surfaces.

However, the situation may be different. With a decrease in the intensification of the welding mode, a transition occurs from the wavy interface considered above to a flat surface observed near the lower boundary of the 'weldability window'. In this case, as was shown above in section 4.4, the surface is covered with splashes, which are shown in Fig. 82. But when passing between the surfaces of the specified type, an intermediate type surface can be observed, which can be called a quasi-wavy surface. It is a piecewise wave surface containing both waves and splashes. Section 7.3.1 is devoted to a detailed study of the quasi-wavy surface in the copper–tantalum welded joint. Figure 107 *a* shows SEM images of splashes, and in Fig. 107 *b...d* the quasi-wavy surface. It can be seen that such a surface is non-uniform and consists of disoriented regions with their own wave directions, their lengths and amplitudes. Most of all, the quasi-wave surface is like a patchwork quilt.

When operating a chemical reactor, its walls are exposed to corrosive media at relatively high temperatures. In this case, instability of the quasi-wavy structure may appear and lead to the destruction of the welded joint.

Thus, the zone of risk for the shell of a chemical reactor is the region belonging to the copper–tantalum weld, which has a quasi-wavy structure.

5.2. Petrochemical reactor (coke oven)

Section 3.5 presents the results of an electron microscopic study of the structure of the transition zone [41] with the parameters of the welding mode used to obtain the shell of the coke oven chamber (Fig. 91). The structure of the transition zone turned out to be complex [69, 70]: it consists of five bands with different nature (Figs. 66, 67).

Fig. 91. Petrochemical reactor (coke oven chamber): *a* – reactor vessel, *b* – risk zones (2 – melt zone; 3 – rod-like eutectoid carbides zone)

It is considered established that the typical values of the diffusion coefficient D at a sufficiently high temperature are about 10^{-10} cm^2/s for the solid phase and 10^{-5}–10^{-4} cm^2/s for the liquid. For carbon, the corresponding diffusion coefficients in steel according to various literature data are approximately $0.8 \cdot 10^{-10}$ cm^2/s and 10^{-7} cm^2/s for the solid phase at elevated and high temperatures and $\sim 10^{-5}$ cm^2/s for the melt [71]. But then carbon segregation will occur at the boundary of the melt and chromium steel, i.e. an area with a higher carbon concentration compared to both steels [41, 72].

The sequence of events that unfold during the formation of such a structure is traced further. We are trying to find out which bands are areas of risk. First of all, pay attention to the concentration heterogeneities of chromium (see Table 8). Such concentration inhomogeneities from the initial concentration of the order of 1 wt. % to concentrations, an order of magnitude greater, are the result of the dispersion of chromium steel particles. As we said above, it is the formation and scattering of particles that first provoke melting. The friction between the particles is sharply increased due to the fact that they have a much larger surface than the surface of the contacting plates and fly away not in free, but in a closed space. But then, due to the high diffusion mobility of carbon in a liquid, a stream of carbon atoms arises from the melt to the solid phase. In this case, it is the flow from area 2 of 12KhM phase to area 3 of steel 08Kh13. At first glance, it could be expected that the concentration of carbon in region 3 would be close to the average concentration in steel 12KhM. But the situation turned out to be much more complicated.

For the compound under study, the segregation zone is indeed observed and has a sufficient width, albeit a smaller one in comparison with the neighbouring zones. The boundaries of the segregation zone are also visible in Fig. 66, and especially clearly in Fig. 63: sharp border – with solidified melt (zone 2), blurred – with zone 4.

The phenomenon of segregation has long been known. There are numerous observations of segregation in a variety of objects: grain boundaries, domains, phase boundaries. But segregation during explosive welding at the boundaries of the melt and the solid phase is observed for the first time. When welding two simple metals, segregation will not occur. Segregation zones can occur only when, in addition to the original two elements, there is a third element. Segregation will not occur if there is no melt zone. This means that of the two materials being welded, the initial concentration of the third element should be higher in the low-melting material, in which during the welding process one can expect the appearance of local melting areas.

For the above mentioned copper–tantalum, aluminium–tantalum welded joints, in which there are melt zones, but there is no third element, segregation does not occur at the boundaries of the melt with the solid phase. In steel–steel, steel–metal welded joints, carbon segregation zones will appear, similar to those observed in the compound under study, if the conditions stated above are fulfilled. Various situations arise for known coatings of steel with titanium or aluminium. The melting point of steel is approximately 1400...1500°C, titanium 1660°C, aluminium 660°C. Therefore, in the case of a steel–titanium welded joint, segregation can be expected due to the flow of carbon from the zone of molten steel to titanium. On the contrary, in the case of a steel–aluminium welded joint, the aluminium melt zone does not contain carbon and will not segregate.

The driving force for segregation is a decrease in the carbon energy in this zone, but not due to the large concentration of the randomly distributed third element, as is often the case during segregation, but due to the transformations that this element causes [73]. For the compound studied here, we are talking about binding large concentrations of chromium and carbon into carbides, having the form of plates or rods and organized in colonies. Thus, the advancement of the interface is accompanied by eutectoid decay with the formation of carbide colonies and a depleted zone. The segregation zone is either formed immediately, and then eutectoid decay occurs, or the zone boundary moves gradually with constant pumping of carbon from the melt region.

Figure 92 shows the concentration of carbon depending on the composition for various carbides [74]. As can be seen from Fig. 92 *a*, the carbon concentration c_{car} in carbides (Fe, $Cr)_{23}C_6$ is about 5.7 wt.% Using micrographs shown in Figs. 64, 65 in Section 3.5, it is possible to estimate, although rather roughly, the average concentration of carbon c_{segr} in the colony. We use for evaluating the cross sections of the colonies and determine the average distance between the rods by the number of their outputs on the section plane. In this way, the dislocation density on a plane perpendicular to their axis is usually estimated by the dislocation density and the distance between them.

We introduce the notation d for the rod thickness and D for the distance between them. The average value of the ratio $d/D \sim 1/3$. Since $c_{segr} \sim c_{car}\, d^2/D^2$, then $c_{segr} \sim 0.6\%$. This is the upper limit of the carbon concentration c_{segr} in the colony. Since the rods are not equidistant, and the colonies are surrounded by a BCC carbon-depleted phase, it can be assumed that the average carbon concentration in zone 3 is lower than the estimate given, but in any case higher than in the original chromium steel. This is to some extent confirmed by the observed in Fig. 66 by the ratio of the widths of zone 3 and zone 2. It is precisely by pumping carbon from zone 2, where the carbon concentration is 0.10...0.12%, that a segregation zone is formed, containing colonies of rod-shaped carbides.

The width of zone 3 is approximately three times smaller than the width of zone 2. Similarly, for carbide (Fe, $Cr)_7C_3$, for which, according to Fig. 92 *a*, the concentration c_{car} is about 9.2%, the

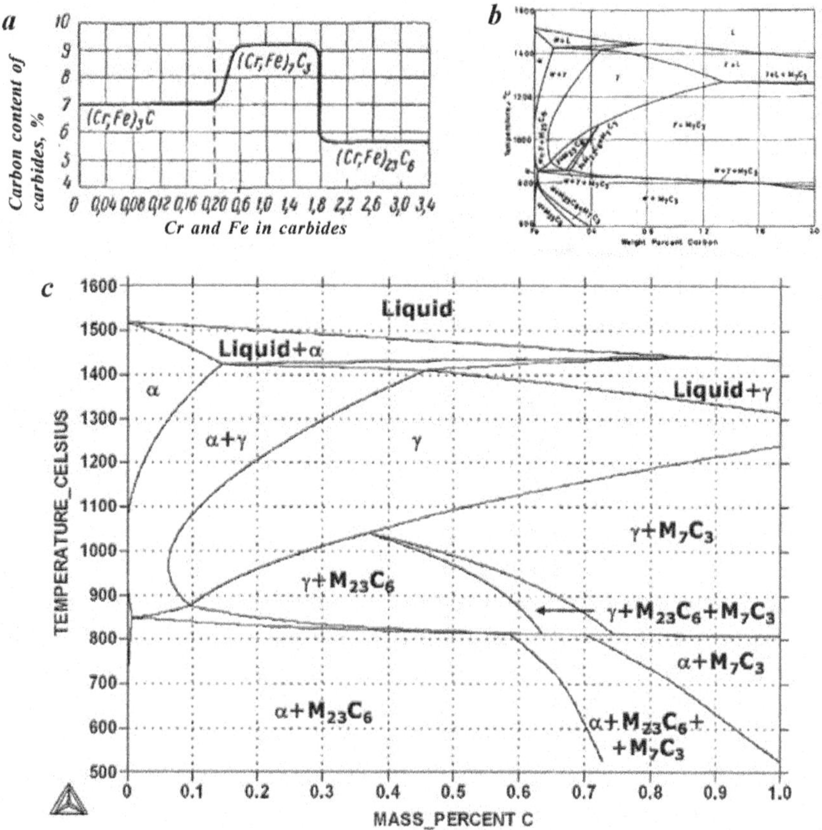

Fig. 92. The carbon content in carbides depending on the concentration (Fe–Cr) (*a*); phase diagram: for steel (C–Cr–Fe) with 13% chromium (b); for stainless steel (Cr 12.8%, Mn 0.65%, Si 0.4%) (*c*).

value of c_{segr} is ~1% (all concentrations are here in wt.%). Figure 92 *b* shows the phase diagram [75] for C–Cr–Fe at 13% chromium. Figure 92 *c* shows the phase diagram [76] for stainless steel (Cr 12.8%, Mn 0.65%, Cu 0.4%), the composition of which is close to that used in this work chromium steel (Cr 13%, Mn 0.8%, Si 0.4%) . Both diagrams are obtained as a result of computer calculations. From Fig. 92 *b* it can be seen that for the first alloy, in equilibrium states, at temperatures below 800°C, at carbon concentrations of 0.3–0.4%, carbides (Fe, Cr)$_{23}$C$_6$ and (Fe, Cr)$_7$C$_3$ are observed, and at 0.4–0.6%, only (Fe, Cr)$_7$C$_3$. For stainless steel, as can be seen from Fig. 92 *c*, at the same temperatures and carbon concentrations only carbides (Fe, Cr)$_{23}$C$_6$ are observed. Note that in both cases there is no carbide (Fe, Cr)$_3$C.

Using SEM, EBSD analysis of the structure of zone 3 was carried out. Three types of carbides were observed. Cementite $(Fe, Cr)_3C$ enriched with chromium is observed along the grain boundaries. Grains are formed by ferrite (α-Fe) containing special carbides $(Fe, Cr)_{23}C_6$ and $(Fe, Cr)_7C_3$. If we compare the content of special carbides, then the greatest proportion falls on carbide $(Fe, Cr)_{23}C_6$. The phase distribution is non-uniform. With distance from the melt zone, the ratio between the phases changes, namely, the content of ferrite increases. In this case, the structure approaches that observed in zone 4. There is a difference with the phase diagrams shown in Fig. 92, where one of the special carbides is observed in the carbon concentration range under study, and this is different in different diagrams. We also pay attention to the fact that for chromium white cast irons, besides cementite $(Fe, Cr)_3C$, eutectoid carbide $(Fe, Cr)_7C_3$, having a pencil form [77], was found.

The possible reason for the discrepancy with the diagrams shown in Fig. 92 *b*, consists in that the phase diagrams give equilibrium phases, and in the case studied here the phases are non-equilibrium. The observed eutectoid carbides have a complex prehistory: during an explosion – ultrafast heating of steel, then quenching, then slow cooling at residual temperatures, then heating to 700°C and tempering at this temperature for two hours, then slow cooling. The resulting structure can be very different from the equilibrium and can be determined by the kinetics of phase transformations, namely, which of the possible, including metastable, phases grows faster. At the moment, we do not have data to determine which of the observed eutectoid carbides has a rod-like shape and is it possible for both carbides simultaneously.

Measurements of microhardness in various bands of the transition zone showed that the microhardness of the rod-like carbide zone (zone 3) reaches a value of 3600 MPa, which is about 2000 MPa higher than the microhardness of recrystallized steel (region 1, 5). Such a high value is due not only to the microhardness of the carbides, but also to the fact that they are combined into colonies. As a result of the appearance of the colonies, the deformation behaviour of the material, due to the action of high and long-term loads, may change significantly, including, possibly, in the direction of reducing the long-term strength. Indeed, during operation the walls of the reactor chamber are subjected to pressure due to the mixing of heavy oil fractions. In this case, colonies of rod-like carbides act as barriers that inhibit the dislocation from sliding in the plane crossing the colony. Due to the fact that inclusions have the shape of rods, such a mechanism for overcoming obstacles by dislocations, such as rounding them, is practically excluded. A mechanism for cutting inclusions by dislocations would be possible for individual inclusions. However, this mechanism is extremely difficult in the case of a colony. The dislocation appears to fall into the 'maze', from which it can not get out. As a result, stress concentrators arise that initiate the development of cracks.

The above mentioned sequence of events that lead to the occurrence of risk zones can be represented as follows: scattering of chromium steel particles, local melting of 12KhM steel, carbon transport through the liquid phase, formation of a segregation zone, eutectoid decomposition of the segregation zone. As a result, a segregation zone containing colonies of rod-shaped carbides is a risk zone for the shell of a petrochemical reactor. But in reality there are two risk zones. The formation of the segregation zone is initiated by the transport of carbon from the local melting zone. As a result, both zones (both melting and segregation) turned out to be risk zones, which are shown in Fig. 91.

Based on the research conducted in this work, we believe that the choice of heat treatment mode (annealing at 680–700°C for about 2 hours) was not the most successful. Its parameters (temperature, duration) turned out to be such that they can provide eutectoid decay. In order to verify this, we needed samples of the welded joint without heat treatment. Comparison of the structure of these samples with their structure after heat treatment showed unequivocally that rod-shaped carbides arose during heat treatment.

Figure 93 *b* shows the melt zone (white arrow) and its adjacent chromium steel strip (red arrow). Rod-like carbides are clearly visible in this band. In addition, the initial sample was heated at a temperature of 500°C for 2 h and 10 h (Figs. 93 c and 93 *d*, respectively). The melting zone, which arose during welding, is visible, However, in the zone of chromium steel next to it, rod-shaped carbides are not observed. As can be seen from the comparison of Figs. 93 *c* and *d*, recrystallization is observed in the melt zone during a longer annealing period. From this it follows that in order to avoid eutectoid decay, heat treatment should be carried out at a temperature of approximately 500°C, ensuring stress relief due to long-term annealing.

Thus, for the shell of a petrochemical reactor, there are two risk zones: the melt zone and the rod-like carbide zone.

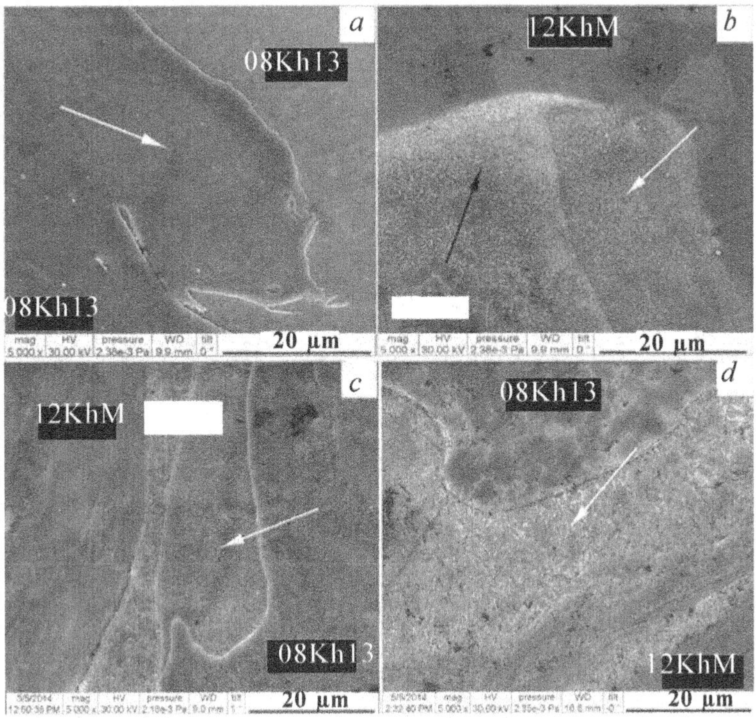

Fig. 93. The melt zone and the adjacent band of chromium melt steel after explosive welding: *a* – before heat treatment; *b* – after heat treatment (680...700°C, 2 h), *c* - after heating at 500°C, 2 h; *d* – after heating at 500°C, 10 h.

6

Fractal analysis of the surface relief

This section is very different from the previous ones: it is not tied to any of the specific welded joints, and moreover, to any of the processes that determine their formation [78]. The search for fractals inside the transition zone was initiated by the observation of objects that appear to be self-similar. These are cusps having the form of splashes, dendrites, 'sticky' fingers, particles of solid phases inside local melting zones, etc.

The unusual microstructure of the joints that arise during explosion welding is due to the fact that explosion welding is a high-intensity rapid effect. With all the variety of materials and welding modes, the central problem is the mixing in the transition zone near the interface. The structure of the transition zone, in its physical essence is very complex, includes the topography of the interface (flat or wavy); heterogeneity of the interface (solid cusps and local melting zones); true or colloidal solutions within zones; particles of phases that do not melt inside solutions.

The interface heterogeneity is three-dimensional. Electron microscopic images provide two-dimensional sections of the interface. The longitudinal sections of both cusps and local melting zones are islands surrounded by other phases. Although the particles inside the local melting zones have a different origin, they are also islands. Islands, in turn, are possible fractal objects. The interpenetration of dissimilar materials in the process of explosion welding makes inevitable the emergence of islands of different colors, which is the specificity of the structure of welded joints.

Observation of the islands and the calculation of their fractal dimensions makes it possible to compare different compounds obtained by explosive welding, both for different pairs of source metals and for the same pair, but with different intensity of external influence. The fractal analysis of the transition zone helps to identify not only the structures and the connections between them, but also the connections between the processes. Fractal analysis revealed a link between the two processes: particle scattering and local melting. Such a conclusion was made on the basis of the observation of complexes of islands of different colours (multifractals).

The formation of cusps as a result of diffusionless emissions of one of the welded metals into the other is a typical stochastic process. It is this nature of the process, as well as numerous electron microscopic observations that indicate the recurrence of structures, stimulate the development of a fractal approach for describing the topography of the interface. It is difficult to expect that the cusps appearing at random would be ideal fractals. However, such objects can be considered fractals if at least one of the following conditions is met: they have a rather complex structure on all scales and an increase in the scale does not lead to a simplification of the structure; they are self-similar or approximately self-similar; they have a fractional metric dimension [33, 79]. There are many definitions of fractals. One of them, rather simplified: a fractal is a structure consisting of parts that are in some sense similar to the whole [79]. The fractal dimension indicates the degree of filling of the space with an object or structure.

6.1. Islands

The SEM images (Fig. 94) of the longitudinal section for the copper–tantalum ($\mathbf{C}_p$) welded joints are trichromatic, and the gray zone corresponds to the molten region containing a mixture of metals. The micrograph in Fig. 94 *a* is similar to that given earlier in Figs. 29 *a* and 30 *a*. Here we use this particular micrograph (Fig. 94 *a*), since it contains the largest number of objects, the fractal dimension of which is calculated.

The tricolor image, shown in Fig. 94 *a*, is converted as follows. First, we leave only the longitudinal sections of the cusps of tantalum. These are white islands. Their environment includes areas filled with copper and localized melting zones. The setting of the white islands is considered a black background. The set of white islands for the compound ($\mathbf{C}_p$) obtained in this way and studied further is presented in Fig. 95 *a*. Similarly, gray islands on a black background for the same compound ($\mathbf{C}_p$) were obtained (Fig. 95 *b*).

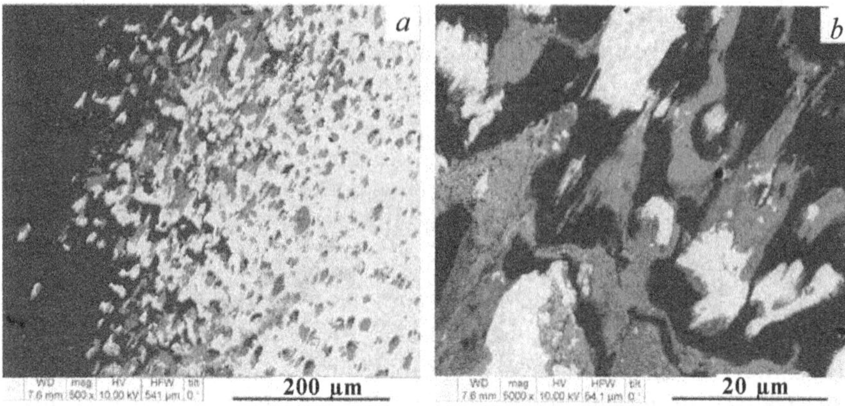

Fig. 94. Cu–Ta (C_p) welded joint, longitudinal section with different magnification (*a, b*); white color — tantalum, black — copper, gray — melting zones, (SEM).

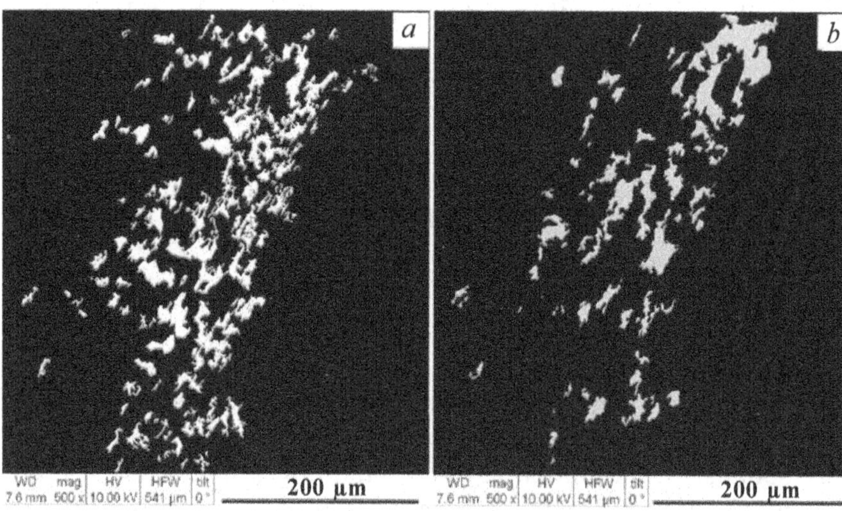

Fig. 95. Islands corresponding to Fig. 94: *a* – tantalum islands; *b* – zones of local melting.

In order to find out whether islands are fractal objects, it is necessary to obtain the ratio between the perimeter of each of the islands and its area. We will carry out calculations of the dimensions of the islands using the methods used in the theory of fractals [33, 79]. Mandelbrot [79] showed that for islands, the outlines of which are somewhat similar, the following relations hold:

$$L(\delta)=C\delta^{(1-D)}[A(\delta)]^{-D/2} \qquad (7)$$

here L is the perimeter, A is the area of the island, D is the fractal dimension, δ is the standard of length.

Relation (7) is satisfied for any standards of length δ, which is small enough to measure the smallest of the islands. Under certain approximations from (7), we obtain a simple proportionality relation, which has the form:

$$A \sim L^{2/D} \tag{8}$$

After logarithmic transformation of (8), we get:

$$\log A \sim \frac{2}{D} \log L.$$

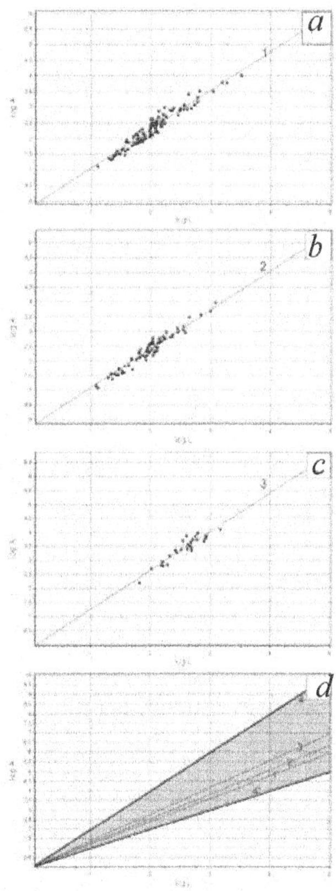

Fig. 96. Cu–Ta welded joint (**C**$_p$), longitudinal section: area–perimeter dependence for tantalum islands (line 1), local melting zones (line 2), tantalum nanoislands (line 3), round islands (line 4): *a, b, c* – the lines built on points (a point - island); *d* – summary schedule.

Area A and perimeter L are calculated for each island. Each island is one point. In Fig. 96 *a* for the white islands, log A is shown to be dependent on log L, which can be approximated by a straight line 1. It is shown how the line 1 is constructed at numerous points using the least squares method. The same line is, to a certain extent, evidence of their fractal nature. Similarly, the measurement results for the gray islands are stacked on the same line 2 (Fig. 96 *b*). Both lines are shown in the summary graph (Fig. 96 *d*).

Knowing the tangent of the angle of inclination, equal to 2/ D, we calculate the value D. Taking into account the error, we obtain: for the islands of tantalum $D_1 = 1.68 \pm 0.07$, for local melting zones $D_2 = 1.58 \pm 0.06$. The procedure was repeated for other longitudinal sections and a fairly good agreement of the dimension values was obtained. However, for some of the cases considered here, the dimension values during the expansion of the base differed from those originally obtained and approached those that will be given below. Line 3 (Fig. 96 *c*) refers to the internal structure of the gray islands and will be discussed below.

Straight line 4 corresponds to the area–perimeter dependence for round islands, when the parameter D takes the smallest value $D_4 = 1$. Line 4 is used to test the program. In Fig. 96 *d*, the line 4 is also represented, which corresponds to the extremely large dimension for two-dimensional fractal objects $D_4 = 2$. As can be seen from the comparison of lines 1, 2 with line 4', for a fixed value of the area A, the corresponding value of the perimeter L is less for the round islands.

This means that an increase in dimension can be achieved due to the irregularity of the islands. As can be seen from the comparison of Fig. 95 *a* and Fig. 95 *b*, the gray islands are less rugged than the white ones. This is because they can blur until they are frozen. As a result, the dimension D_2 of the gray islands is significantly less than the dimension D_1 of the white islands. The high irregularity of the cusps, and accordingly of the tantalum islands, having the largest dimension of those calculated in the present work, helps the cusps most effectively fulfill the role of wedges connecting the contacting materials.

To determine the effect of the mutual solubility of the starting metals, compare the fractal dimension of the islands for the welded joints ($\mathbf{C}_p$) copper–tantalum and ($\mathbf{E}_p$) aluminium–tantalum. For the ($\mathbf{E}_p$) welded joint in Fig. 30 *b* a three-colour image of the interface is shown. One can see the sticking of the islands when approaching

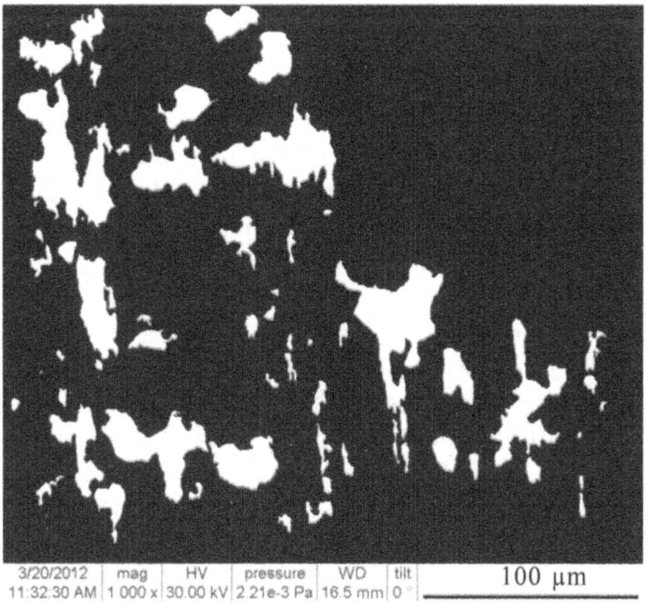

Fig. 97. Al–Ta ($\mathbf{E}_p$) welded joint, longitudinal section (SEM): tantalum islands (white color) on a black background.

the tantalum plate (due to the fact that the cross section is actually not longitudinal, but inclined). In Fig. 97 the tricolour image is transformed in the same way as above: white tantalum islands are highlighted against a black background.

We believe that, in the general case, the convex shape of the islands lowers the fractal dimension, while the presence of concave segments in the outline of the islands, on the contrary, increases the dimension.

The value of D obtained for the aluminium–tantalum welded joint, equal to (1.31 ± 0.06), is much less than the D_1 for the white islands obtained above for the Cu–Ta ($\mathbf{C}_p$) welded joint. It follows that the islands are more irregular for the ($\mathbf{C}_p$) welded joint than for the ($\mathbf{E}_p$) welded joint.

It can be assumed that this is due to the presence of mutual solubility for an aluminium–tantalum pair, due to which the cusps of tantalum in molten aluminium become smoother. Accordingly, tantalum islands are less rugged in this case (Fig. 97). On the contrary, due to the lack of mutual solubility for the copper–tantalum pair, the tantalum cusps, and, accordingly, the tantalum islands (Fig.

94 *a*), in molten copper are strongly cut up. Paradoxically, the lack of mutual solubility, which was considered dangerous, in some cases contributes to weldability.

In both cases, for the ($\mathbf{C}_p$) and ($\mathbf{E}_p$) welded joints, the local melting zones, not being isolated, are pressed to the islands of tantalum. However, the area of these zones for Al–Ta is much larger than for the Cu–Ta welded joints. Perhaps this is due to the low melting point of aluminium (933K).

Islands of different colours inevitably arise when interpenetration of dissimilar materials in the process of explosive welding takes place. An analysis of the distribution of islands on numerous SEM images of the interface showed that in most cases the white and gray islands are combined into complexes. Such complexes are visible in Figs. 29, 30, 94 and especially clearly in Fig. 29 *b* and 94 *b*. At the same time, white islands are seen as isolated very rarely. The calculation of the dimensions of the islands that make up the complexes showed only a slight difference from the values obtained above.

Such complexes are multifractals, which arise as a result of not one construction algorithm, but several successive algorithms [33]. For their description, a multifractal spectrum is calculated, which includes a number of fractal dimensions inherent in the elements of this multifractal. The above dimension values for the white and gray islands can be used as a multifractal spectrum.

Thus, it was found that the main element of the structure are complexes consisting of white and gray islands. For the first time, fractal analysis revealed a link between two processes: particle scattering and local melting. A possible scenario of how the scattering of particles and their inhibition when interacting with an obstacle lead to local melting was discussed in Section 4.5.1. In fact, the scattering of particles binds the gray molten region to the white region of tantalum. This mechanism works regardless of whether the initial phases have mutual solubility.

In addition to the islands, which are the result of a longitudinal section of the cusps (Figs. 95, 97), the islands, which are the longitudinal sections of particles of the refractory phase inside the local melting zones, were investigated. For the titanium–aluminide ($\mathbf{A}_w$) welded joint a local melting zone is shown (|Fig. 20) containing aluminide particles that are immersed in a solid solution of titanium doped with niobium and aluminium. The fragmented layer consisting of non-interlaced aluminide particles is shown in Fig. 18 *c*. The shape of the islands inside the local melting zone for the ($\mathbf{A}_w$) welded joints is clearly visible in Fig. 20 *a*: smooth and elongated. In Fig. 18, the fragmented layer contains not only smooth, but also rugged islands.

It is obvious that the expansion of smooth islands is an easier process than rugged ones. Therefore, smooth islands fly away, while the cut ones remain on the edge of the refractory phase. Indeed, in the ($\mathbf{A}_w$) welded joint, the islands inside the local melting zone have a dimension of approximately (1.32 ± 0.09), while the islands inside the fragmented layer have a dimension of approximately (1.49 ± 0.08).

Local melting zones are filled with either true solutions in the presence of mutual solubility, or colloidal in its absence [52]. For Cu–Ta welded joints, regardless of whether the boundary has a flat ($\mathbf{C}_p$) or wavy ($\mathbf{C}_w$) shape, the local melting zone is a frozen colloidal solution of two immiscible phases. Due to the high melting point of tantalum (3300 K), the colloidal solution consists of a copper melt containing particles that have not experienced the melting of tantalum. Upon further solidification, the colloidal solution becomes a frozen dispersion-strengthened suspension.

For the ($\mathbf{C}_p$) welded joint the bright-field TEM image, shows a lot of dark tantalum particles of arbitrary shape (Figs. 32, 33). Their sizes are about 30...50 nm. The points corresponding to each nanoisland with the coordinates of the area–the perimeter are laid on a straight line 3 (Fig. 96 *c*). The dimension of tantalum nanoislands is approximately $D_3 = 1.44 \pm 0.03$, i.e. below the dimension of the islands of tantalum, which are sections of cusps (line 1). But these islands have different origins and varying degrees of irregularity. At the same time, the dimension of the nanoislands still differs from the dimension of the round islands (line 4 in Fig. 96 *d*).

The tantalum islands within the local melting zones are significantly different for the ($\mathbf{C}_p$) and ($\mathbf{C}_w$) welded joints. For the welded joint ($\mathbf{C}_w$), these are micron islands (Fig. 38 *c*). From the comparison of their images one can see that the micron islands are much smoother. The corresponding dimension (1.22 ± 0.09) is the smallest value obtained for the compounds studied here.

Thus, the ideas developed above are fully supported: firstly, the rugged particles have a higher dimension than smooth ones, and secondly, that smooth particles fly away more easily than rugged ones.

The 'smooth–rugged' contrast is a purely qualitative description of the state of the surface. The fractal dimension gives a quantitative characteristic that allows one to simultaneously take into account the area and perimeter of the studied elements of the interface, as well as their changes when the mode changes or when moving from one welded joint to another. To clarify the results, it is necessary to use high resolution transmission electron microscopy (HRTEM). In [80, 81], the effectiveness of this method for investigating the fine structure of interlayers near the interface was shown.

In Figs. 32, 33, *a*, used to calculate the dimension $D_3 \approx 1.44$, chains of tantalum islands are visible. But at higher resolution, these chains can break up into separate nanoislands. As a result, their dimension may fall below ≈ 1.44. However, another option is possible. The density of particles is really so high that they actually stick together. Then the chains will remain at a higher resolution. Accordingly, the high value of the dimension will remain.

A fractal analysis of the islands allowed us to suggest that the cusps on a flat border, unlike what was previously thought [11], do not have the shape of cones. Indeed, for the $(\mathbf{C}_p)$ and $(\mathbf{E}_p)$ welded joints, the white islands have an irregular shape, which the sections of the cones cannot have. It is the longitudinal sections of such cusps that are investigated above the islands. If the second element is etched, then the SEM images of tantalum cusps for the $(\mathbf{C}_p)$ welded joints are shown in Fig. 82, for the $(\mathbf{E}_p)$ welded joints – in Fig. 84. With all the angles used, cusps have the form of splashes. An essential feature of the image of the cusps is the repeatability of self-similar relief elements (Fig. 83), which is evidence of their fractal nature.

On the consolidated Fig. 98 for the copper–tantalum $(\mathbf{C}_p)$ welded joints, the area–perimeter lines are given for white and gray islands resulting from a longitudinal section of tantalum cusps and local melting zones, respectively, as well as tantalum nanoparticles inside local melting zones. Here are micrographs of the islands.

In Fig. 99 is a map of the coast of Norway [33], for which Feder calculated the dimension of the coastline. In a similar way, it is possible to calculate the dimension of the coastline observed with a longitudinal section of the wavy border of the welded joint. In contrast to the flat boundary investigated above, the longitudinal section of the wavy boundary is a set of alternating bands of both raw materials.

Fig. 99 is a map of the coast of Norway [33], for which Feder calculated the dimension of the coastline.Similarly, it is possible to calculate the dimension of the coastline observed with a longitudinal

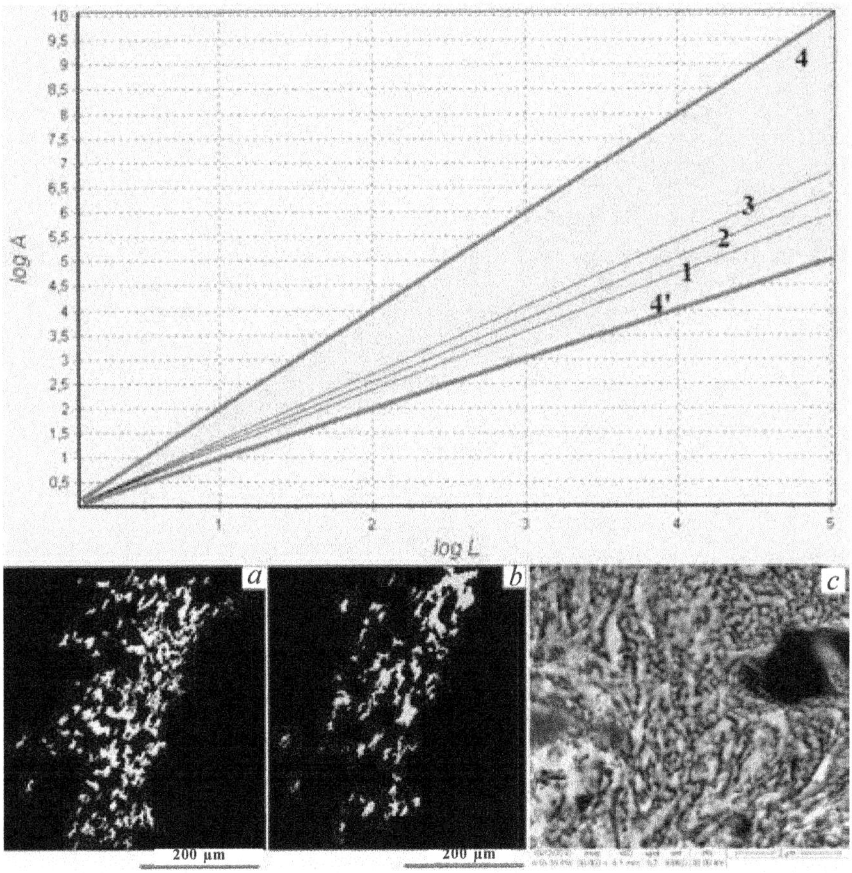

Fig. 98. Copper–tantalum (C_p) welded joint: 1) white islands; 2) gray islands; 3) white islands inside the local melting zones; 4) round islands.

6.2. Coastline

section of the wavy border of the welded joint. In contrast to the flat boundary investigated above, the longitudinal section of the wave-shaped boundary is a set of alternating bands of both raw materials. There are many beaches in contrast to the coast in Fig. 99. In addition, in Fig. 99 territories of two colours (sea – land) are shown, whereas in Fig. 100, and taking into account local melting zones (bays) – three colours.

The dimension of the coastline was calculated in the simplest way. From the set of squares with side δ, we determine the number of those that cover the coastline. As a result, we obtain the dependence

Fig. 99. Map of the coastline of Norway [33].

$N(\delta)$. We use the relation that is performed asymptotically for small δ:

$$N(\delta) \sim \frac{1}{\delta^D} \tag{9}$$

From (9) we determine the fractal dimension of the coastline. In order to avoid the third color when calculating the dimension, local melting zones are included either in black or in white stripes. To calculate the dimension of the coastline in Fig. 100 it is necessary to retain the territory of two colours.

Figures 100 *b...d* show the elements of the landscape corresponding to Fig. 100 *a*. At the same time, local melting zones (bays), as already mentioned, are included in black copper bands. Using formula (9), we determine the fractal dimension of the coastline for the compound under study. We start the iteration, decreasing δ with each next step, and each time we get all new $N(\delta)$.

Figure 101 illustrates this mechanism: from the largest value of δ in Fig. 101, and to the smallest in Fig. 101 *d*. Next, we build a graph of the dependence of log $N(\delta)$ on log δ. Each iteration is one point on

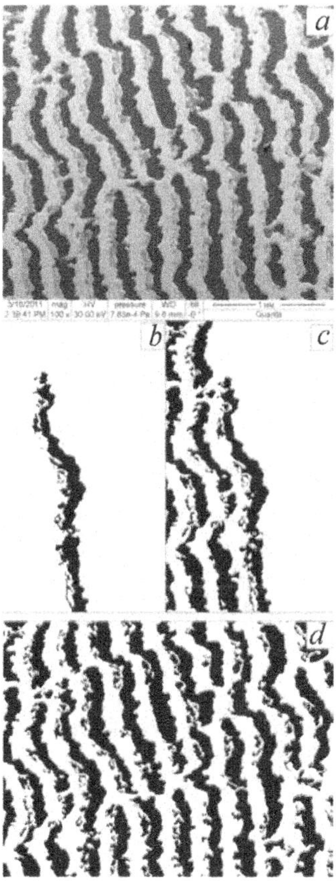

Fig. 100. (C_w) welded joint Cu–Ta, longitudinal section: *a* – SEM image; *b*...*d* – a different number of bands, local melting zones are included in black bars.

the graph. Then we find the angular coefficient of the approximating straight line. Next, we determine the fractal dimension.

In Fig. 102, a similar graph is shown as an example. The lower straight line, with the index "1", corresponds to one band (Fig. 100 *b*), and has a fractal dimension $D = 1.19 \pm 0.03$. The upper one with the index "2" corresponds to several bands (Fig. 100 *c*), which reach saturation, and has a fractal dimension $D = 1.31 \pm 0.02$. A further increase in the number of bands does not lead to a change in the fractal dimension.

For the coastline of Norway, $D = 1.52$ [33], We used the map of the coastline of Norway, shown in Fig. 100, for testing our program, with the help of which the D value was obtained, which is rather close to the value of 1.52.

Thus, for the studied welded joint (**C**$_w$), copper–tantalum, having a wavy border, the fractal landscape includes numerous channels with indented banks, as well as bays containing numerous islands.

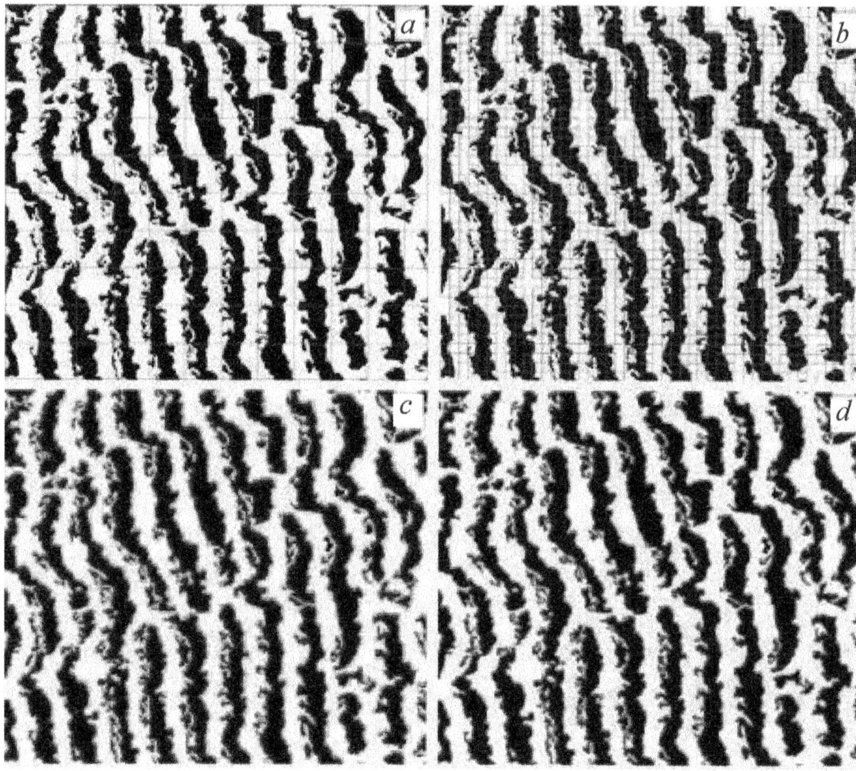

Fig. 101. Cu–Ta (C_w) welded joint, wavy border. The method for determining the fractal dimension.

Conclusions to chapter 6

1. A new fractal approach is proposed to describe the structure of the transition zone of welded joints. Various types of fractals were found: islands, multifractals, coastline, and their dimensions were calculated.

2. The islands are the result of a longitudinal section. surface irregularities, such as cusps and local melting zones. It is the islands of different colours that inevitably arise when interpenetration of dissimilar materials in the process of explosion welding.

3. For a copper–tantalum compound with a flat boundary, the largest fractal dimension ($\approx 1.68 \pm 0.07$) has white islands (sections of tantalum cusps). This is due to their greater irregularity. In accordance with this, the surface of the cusps itself has a large irregularity. This is the special role of the cusps as 'wedges' that keep the welded joint from fracture.

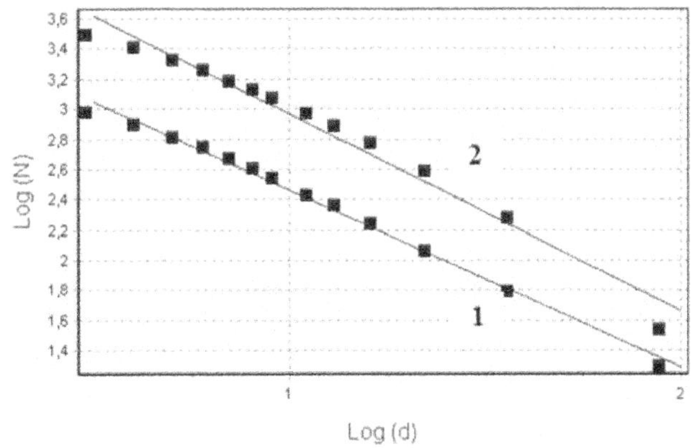

Fig. 102. Dependence of coastline length on cell length, corresponding to Fig. 100, *b* (line 1) and Fig. 100 *d* (line 2).

4. For the first time, fractal analysis revealed a link between two processes: particle scattering and local melting. It is the formation and dispersion of particles that, first of all, provoke local melting due to friction between the particles and their environment. This is confirmed by the observation of complexes consisting of islands of different colours (multifractals).

7

Evolution of the interface of copper–tantalum and aluminium–tantalum welded joints

To identify the main processes that determine the weldability, it is necessary to study the structure of welded joints. In this chapter, the field of electron-microscopic study of welded joints is expanded by including the evolution of the structure in it as the welding mode intensifies. Special attention is paid to the analysis of inhomogeneities of various types inherent in the interface: at the lower boundary (LB) of the window of weldability, near LB, both below and somewhat above the boundary, and also inside the window of weldability.

The structure of the interface is determined, in our view, by two types of heterogeneity: cusps and local melting zones. The nature and mechanisms of the very transient appearance of cusps at the interface of materials during explosion welding have not yet been fully elucidated. Nevertheless, it can be assumed that the cusps arise as a result of diffusionless (due to the transience of welding) ejection of one metal into another.

The research carried out below is aimed at identifying the characteristic structure of the interface, which must be achieved in order for welding to occur. There is a whole range of issues that this chapter is devoted to:

- what shape do the cusps have below LB;
- what shape the cusps have near LB;
- which relief has an interface above LB;

- what is the evolution of the relief of the interface with the intensification of the welding mode.

Accordingly, the study is conducted in the following directions: cusps at different angles below LB, turning cusps into splashes when passing through LB, converting splashes into waves, imperfect and perfect structure of the wavy interface inside the weldability window.

7.1. Material and research methods

Explosive welding was carried out by the Volgograd State Technical University, OJSC Ural Chemical Engineering Plant (Yekaterinburg). Copper and tantalum, which do not have mutual solubility, were selected as the main source materials. This chapter continues the study of compounds ($\mathbf{C}_p$), ($\mathbf{C}_w$), conducted above (see Section 3.2). The compound ($\mathbf{C}_p$) was obtained near the lower boundary of weldability, the compound ($\mathbf{C}_w$) inside the weldability window. Welding was also carried out with parameters below LB. This mode is conventionally denoted as ($\mathbf{C}_p\downarrow$). When using the above-mentioned intermediate modes, special attention is paid to the transitions through the lower boundary (LB) of the "weldability window" and to clarify the reasons why weldability is impossible below LB, although setting is possible. The lower boundary of the 'weldability window' is important both for practical calculations of welding conditions and for understanding the processes that determine the possibility of forming a welded joint. Welding modes near LB are characterized by minimal values of parameters. In addition, the modes (a), (b), (c), (d) were used slightly above LB, for which the border is wavy, but unlike (C_w), is extremely heterogeneous.

In general, the modes used as they intensify are arranged as follows:

($\mathbf{C}_p\downarrow$) $\gamma = 3.8°$, $V_{con} = 2564$ m/s
($\mathbf{C}_p$) $\gamma = 5.22°$, $V_{con} = 2571$ m/s
$\mathbf{C}_w^{(a)}$, $\gamma = 8.6°$, $V_{con} = 2000$ m/s
$\mathbf{C}_w^{(b)}$, $\gamma = 8.9°$, $V_{con} = 2064$ m/s
$\mathbf{C}_w^{(c)}$, $\gamma = 10°$, $V_{con} = 2069$ m/s
$\mathbf{C}_w^{(d)}$, $\gamma = 10.8°$, $V_{con} = 2074$ m/s
($\mathbf{C}_w$), $\gamma = 11.8°$, $V_{con} = 2140$ m/s

The totality of the experimental data obtained forms the basis of new ideas about wave formation [82, 83].

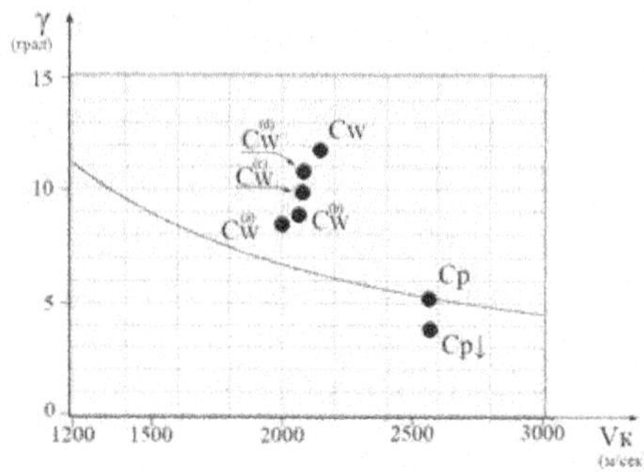

Fig. 103. Parameters of the modes used (γ – impact angle; V_{con} – contact point speed).

7.2. Relief of the flat surface section

When using the above-mentioned intermediate modes, special attention is paid to transitions through the lower boundary (LB) of the 'weldability window' and clarification of the reasons why weldability is impossible below LB, although setting is possible. As mentioned in Chapter 2, the lower boundary is important for understanding the processes that determine the possibility of a welded joint formation.

The research carried out in this chapter is aimed at identifying the characteristic structure of the interface, which must be achieved in order for welding to occur. To find out which relief the tantalum surface has for copper-tantalum compounds after explosion welding, the procedure described in Section 4.4 was used. Copper was etched. Due to the high corrosion resistance of tantalum, the local melting zone also disappeared. Then, SEM images of the tantalum surface were obtained with its different orientation relative to the incident beam, i.e. for different values of characteristic angles (see Fig. 82 *a*).

7.2.1. ($C_{p\downarrow}$) copper–tantalum welded joint, below the lower boundary

Below yhe lower boundary of the weldability the adhesion of the plates in the general case does not occur. However, it turned out that for the mode (**$C_{p\downarrow}$**) on a certain area, welding nevertheless occurred.

In Fig. 104 *a* the upper left corner is the area where the Cu–Ta welded joint appeared (see insert). On the rest of the area, the welded joint, at first glance, also formed. But when polishing the

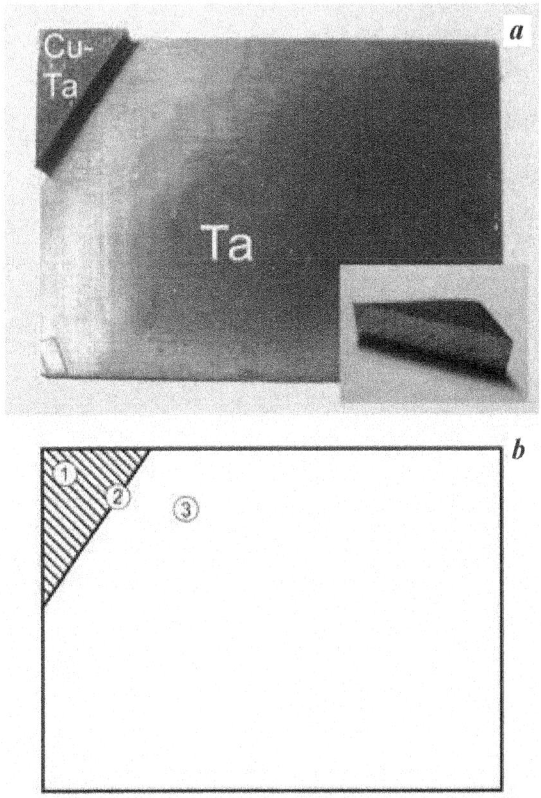

Fig. 104. The surface of tantalum is below LB, the compound (**C**$_{p\downarrow}$), sample size 4×6 (cm): *a* – the area where welding took place; *b* – schematic illustration.

sample, the welded joint stratified with the exception of the triangle mentioned above. Fig. 104 *b* schematically shows the shaded area I, where welding took place.

For area I, an SEM image of a longitudinal section was obtained before copper was etched. The image turned out to be tricolour, similar to what is shown in Fig. 30. After copper was etched, splashes were observed on the surface of tantalum (Fig. 105 *a*), similar to those in Fig. 82. Moving from area 3 towards hatched area 1, one can observe a change in the surface relief of tantalum: in the border area 2: the cusps become larger, and the width of the bands separating them decreases (Fig. 105 *b*). Here, the specified width is not more than 100 μm, whereas in region 3 approximately 300...400 μm. At the same time, both in area 3 and area 2, alignment of the cusps in rows is seen. However, such a number of cusps is not enough to ensure weldability, but may be sufficient for setting.

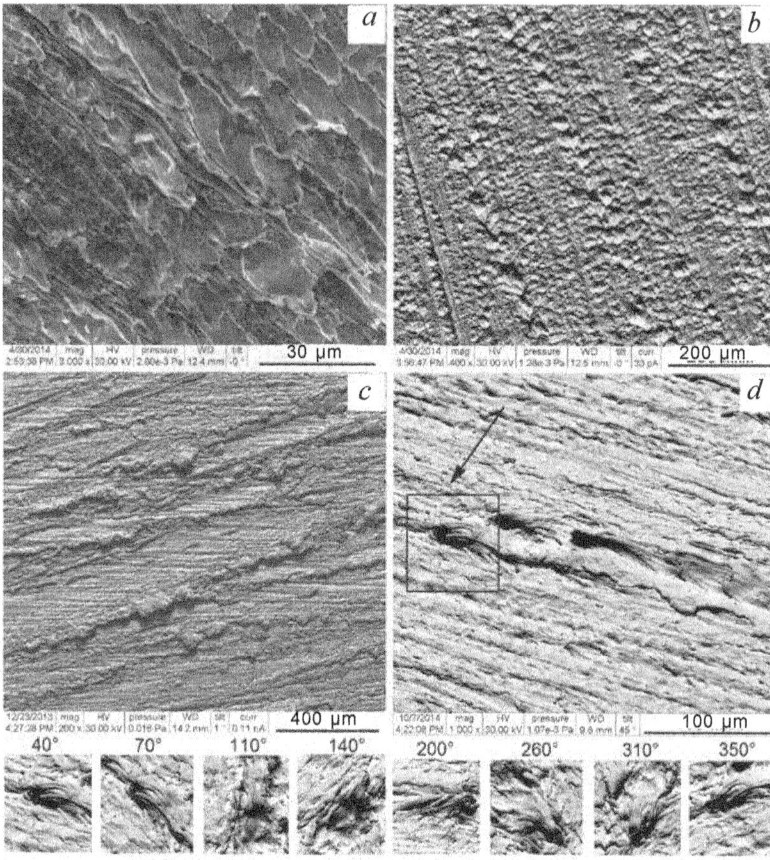

Fig. 105. SEM image of the tantalum surface below LB, welded joint ($\mathbf{C}_{p\downarrow}$): *a* – region 1; *b* – area 2; *c, d* – area 3, different aspects.

Figure 105 *d* shows a set of SEM images obtained at different values of characteristic angles, i.e. at different aspects [82]. It was possible to fix the same big ledge and look at it from all sides.

In order to describe the overall topography of the interface between different materials, it is necessary from a variety of angles, giving two-dimensional images, to reconstruct 3D objects on this surface, which are the cusps [84...86]. The problem of reconstructing three-dimensional objects from their perspectives is solved, as is well known, in a wide variety of fields, including medicine, architecture, painting, criminology, etc. Thus, although the welding parameters in this case (variant ($\mathbf{C}_{p\downarrow}$)) are outside the weldability window, they were in a separate area at the lower weldability limit. In this case, on the same sample, one can observe the transition from the area where welding is observed to the area where it did not occur. Here, the non-uniformity of high-intensity force during explosive welding is manifested, which implies a large plastic deformation, friction of surfaces, the influence of a cumulative jet and other factors.

7.2.2. ($E_{p\downarrow}$) aluminium-tantalum below the lower boundary

Below the lower boundary welding did not occur. In this case, conventionally designated as ($\mathbf{E}_{p\downarrow}$), the surface relief of tantalum has the form shown in Fig. 106.

The similarity of the images in fig. 106 for compounds aluminium - tantalum and in fig. 105, b, c for copper – tantalum compounds. Thus, in general, the surface structure below LB for these compounds turned out to be almost identical, despite the significant difference in the mutual solubility of the starting metals.

7.2.3. (C_p) copper - tantalum at the lower boundary

Figure 82 *b...d* shows clearly the regular distribution of characteristic cusps for the longitudinal section of the transition zone. One can clearly see the alignment of the splashes in rows along the selected direction. This direction is parallel to the line of intersection of the planes of the missile and fixed plates. The distance between the rows is approximately 30–50 μm. The images given in the monograph are the first observations of cusps in the form of splashes. On all images

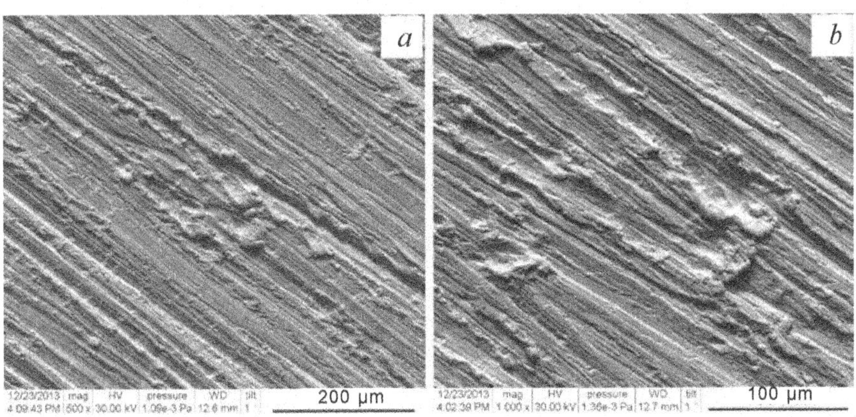

Fig. 106. SEM image of the tantalum surface for the ($\mathbf{E}_{p\downarrow}$) welded joint (aluminium etched): *a, b* – below LB at different magnifications.

obtained at different angles, dense groups of splashes in contact with each other are visible in Fig. 107, and they are marked by arrows.

Figure 27 *c* shows the cross section of the transition zone for the welded joint (C_p). It is clearly seen that the interface is not smooth, but contains cusps. The dimensions of the cusps are approximately 5...10 μm. For comparison Fig. 27 *d* is an SEM image of the surface of tantalum prepared for welding. The initial roughness is 10...20 times smaller than the size of the cusps.

7.3. Relief of the wavy interface

7.3.1. ($C_w^{(a)}$), ($C_w^{(b)}$) copper – tantalum welded joints near (above) the lower boundary

This section presents the results that were obtained using the alternative to the previous parameter changes, namely, near the lower border of the weldability window, but slightly above it (modes (*a*) and (*b*)). Accordingly, welded joints are designated as ($C_w^{(a)}$ and $C_w^{(b)}$). For the $C_w^{(a)}$ weld Fig. 107 *b* shows a wavy but very heterogeneous surface: in different areas different wavelengths and amplitudes.

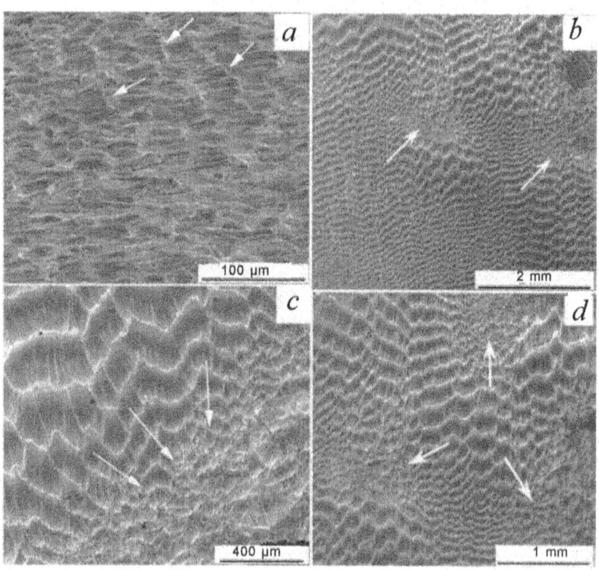

Fig. 107. SEM image of the surface of tantalum for copper–tantalum welded joints (copper etched): a - (C_p) weld, splashes; *b*...*d* - connection ($C_w^{(a)}$), waves and splashes (indicated by arrows) at different magnifications.

Dangling crests of tantalum are visible.

Such a surface can be called a quasi-wave. In addition, in some places there are also splashes. They are indicated by arrows in Fig. 107 *c, d*. Here, for the first time, splashes and waves were also observed simultaneously. Figure 39 shows the image of the wave-like surface of tantalum for the ($\mathbf{C}_w$) welded joint (copper etched). The regular distribution of the tantalum crests contrasts sharply with the chaotic distribution observed in Fig. 107 for a quasi-wave surface.

Most of all, the quasi-wave surface is similar to a 'patchwork' quilt consisting of parts with its direction of the tantalum crests, its wave parameters, and its splashes at the borders of the parts. This unusual surface has not previously been observed.

A similar picture was obtained for mode (*b*), slightly different from mode (*a*), but somewhat more rigid. Figure 108 shows structures

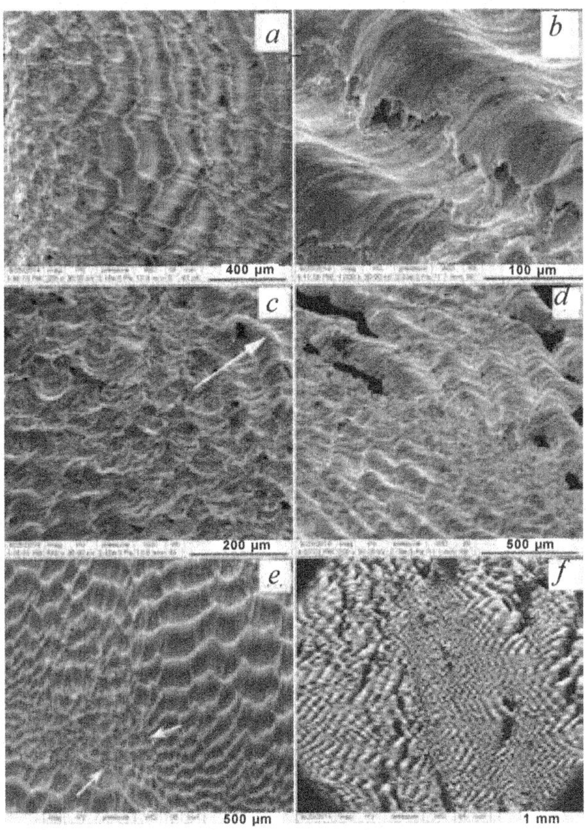

Fig. 108. Welded joint $\mathbf{C}_w^{(b)}$: *a...f* – waves and splashes at different magnifications.

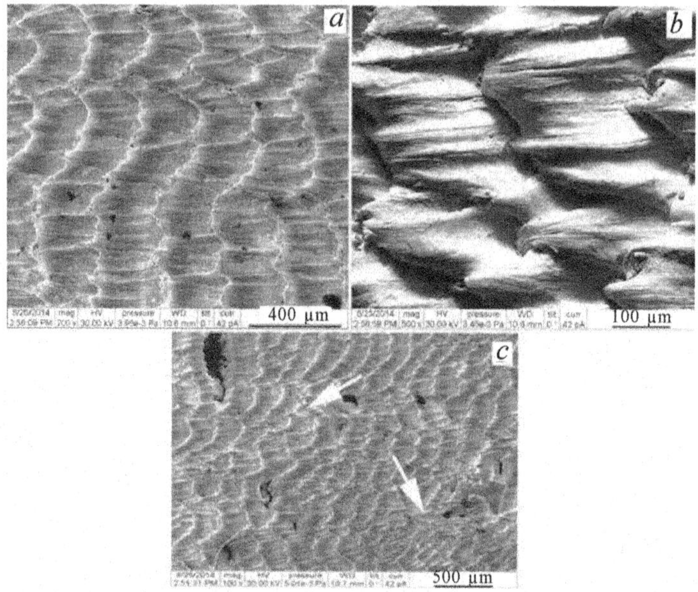

Fig. 109. ($C_w^{(c)}$) welded joint: *a, b* – waves with different magnification; *c* – splashes and waves.

of the interface for the ($C_w^{(b)}$) welded joint: wave-like surface (Fig. 108 a, b), splashes (Fig. 108 *c*), waves and splashes (Fig. 108 *d, e*), 'patchwork quilt' (Fig. 108, *f*).

Similar images of a non-uniform interface were observed for both welded joints ($C_w^{(a)}$) and ($C_w^{(b)}$). But in the case of a welded joint ($C_w^{(b)}$), in contrast to a welded joint ($C_w^{(a)}$), the wavy surface (Fig. 108, *a*) occupies a considerable area. Let's pay attention to the image of a wave received with the big increase on Fig. 108 *b*. Although the waves occur in the solid phase, but this image resembles a wave on the water. Splashes in Fig. 108 *c* are similar to those observed for the (C_p) welded joint (Fig. 82). Here also, as in Fig. 107 *a*, visible group of splashes is pressed to each other (shown by the arrow).

In this case, the merging of splashes into groups occurs above LB, where waves should already form. This indicates a related relationship between these processes. In Figs. 108 *d, e* such large groups of splashes are visible when the line between them and the waves is already deleted. And, finally, the 'patchwork' quilt on Fig. 108 *f* is similar to the one shown in Fig. 107 *b*, but not identical to it.

7.3.2. ($C_w^{(c)}$), ($C_w^{(d)}$) copper–tantalum welded joint above the lower boundary

With a further intensification of the welding mode and the transition to the ($C_w^{(c)}$), ($C_w^{(d)}$) welded joints the splashes gradually cease to be observed, and the wavy surface becomes more and more perfect. Figure 109 *a, b* shows for the $C_w^{(c)}$ welded joint waves at different magnifications, and in Fig. 109 *c* splashes are visible on the background of the waves.

Fig. 110 *a* shows a wave-like border for the welded joint($C_w^{(b)}$) obtained for mode (*d*) close to the centre of the weldability window. In this case (Fig. 110 *b*), separate groups of splashes are still observed, although much less frequently than for the ($C_w^{(c)}$) weld. The image in Fig. 110 *a* slightly differs from what was obtained for the (C_w) weld at higher welding parameters (Fig. 39). However, the wavy surface in Fig. 39 is more regular. For the (C_w) welded joint, the wavelength and amplitude are approximately ~270...350 μm and ~60...65 μm, respectively [34].

But even in this case, it can be seen that the direction of the crests changes when moving from one area to another. In fig. 39, b, the waists are visible between the tantalum crests, leading to the formation of complex three-dimensional defects that substantially distort the interface surface. It is clearly seen that even for the joint (C_w), not to mention the welded joints discussed above (Figs. 108... 110), the interface does not have an ideal wavy shape. However, it is the (C_w) welded joint inside the chemical reactor wall that ensures

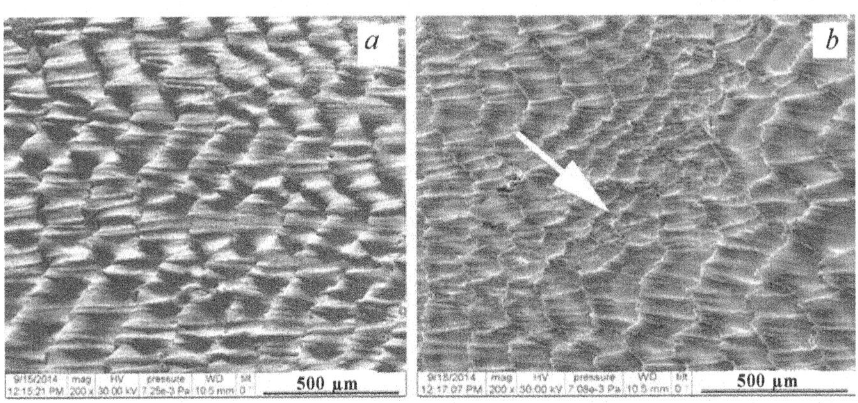

Fig. 110. ($C_w^{(d)}$) welded joint: a – waves, b – waves and splashes.

its high quality and stability under harsh operating conditions (see Section 5.1).

Being precursors of waves, splashes accompany their observation from the lower boundary to the centre of the weldability window. Using the data of the profilometer presented in Fig. 111, one can estimate the magnitude of the excess area associated with the presence of splashes. With rather rough estimates, we find that the excess surface area of the splashes is approximately 10% of the area of the flat boundary. In fact, this is a low estimate. If we take into account that the cusps are not smooth, but rather cut up, the excess area can increase greatly. Due to the large area, the adhesion of the contacting surfaces can be ensured, as is often considered, for example, at the expense of van der Waals forces. However, another reason is possible, due to the fact that the surface of the cusps, as can be seen in Fig. 105 *d* contains concave segments, troughs, cavities and other complex irregularities. In order to ensure the continuity of the connection, they must be filled with another material, both during the explosion itself and during further solidification if there is a melt. Therefore, to uncouple different materials, a rather large force is required, which is necessary for the physical "breaking down" of the cusps. This means that tear resistance will be due to purely topological reasons. In other words, splashes can make a topological connection between the contacting surfaces and thereby ensure strong adhesion between them.

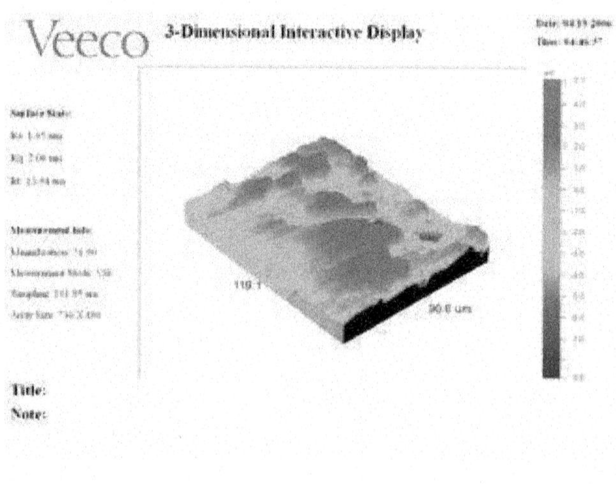

Fig. 111. Profile pattern of a flat interface for a (C_p) welded joint.

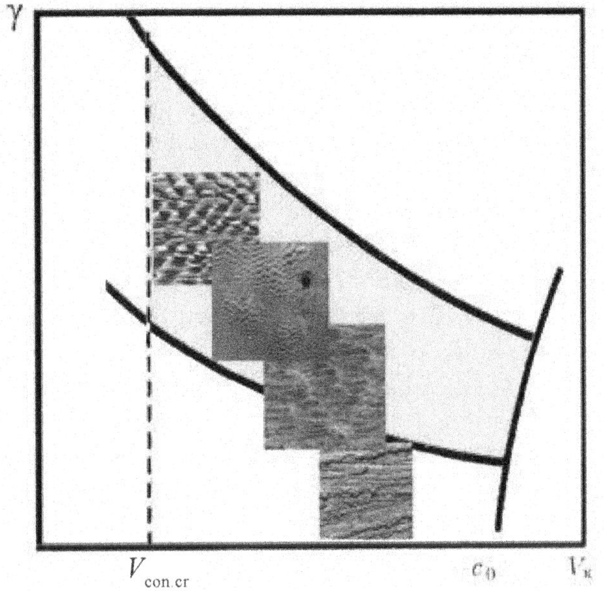

Fig. 112. Typical surface structures for copper–tantalum joints on the background of the weldability window.

The result of the chapter is the identification of a sequence of structural states of the interface, replacing each other as the welding mode intensifies. In Fig. 112 they are depicted against the background of the weldability window. Below the LB (lower boundary), the structure observed in the setting area consists of isolated cusps. Splashes appear on a flat surface near the LB. Slightly higher than the LB, there is a structure called the 'patchwork quilt', which is a piece-wave surface containing both waves and splashes. Inside the weldability window, closer to its centre, a fairly perfect wavy surface is observed.

Conclusions to chapter 7

1. An unusual shape of cusps was found on a flat interface during explosion welding. They are visually similar to splashes that occur on the surface of the water. However, the splashes observed after explosion welding are formed by a solid phase that did not undergo melting.

2. It is shown that below the lower weldability limit, the structure observed in the setting area consists of isolated cusps. An SEM image of the same isolated cusp was obtained using multiple aspects.

3. It was found that near the lower boundary, somewhat higher from it, a quasi-wave surface, very heterogeneous, is observed: in different regions, different wavelengths and amplitudes of waves. In addition, splashes were also found in some places. The sectional surface has an unusual patchwork quilt-like structure.

4. A set of typical interface structures alternating with each other as the regime is intensified is presented against the background of the weldability window. Some of them, such as splashes and 'patchwork', have not been previously observed.

There remain questions related to the extent to which the findings are universal. It is necessary to expand the range of the studied welded joints, including at least those that differ from copper–tantalum welded joint in the type of solubility of the starting metals. The following chapter is devoted to the study of copper–titanium welded joints with limited solubility.

8

Evolution of the interface of copper–titanium welded joints

A new approach to describing the formation of welded joints was proposed on the basis of the study of copper-tantalum welded joints. This approach includes the processes mentioned above: granulating fragmentation and the formation of inhomogeneities at the interface (cusps, local melting zones). The situation is complicated by the fact that with the intensification of the welding mode, the cusps change their shape: at the lower weldability limit, they acquire an unusual shape, similar to splashes on the surface of water, although they are solid. In addition, the situation is complicated, since the transition from a flat interface to a wave-shaped one does not occur suddenly, but through some intermediate states. For copper–tantalum welded joints that do not have mutual solubility, intermediate states have a peculiar structure, which, as can be seen from Fig. 107 *b, d*, 108 *e*, consists of randomly distributed separate regions with their own wave parameters, and these regions are separated from each other by splashes.

The proposed approach allows us to understand and consistently explain the totality of experimental facts obtained for metal welded joints, both having and not having mutual solubility. In the presence of mutual solubility, diffusion and intermetallic reactions are possible. As such an object of study, welded joints copper–titanium metals with limited mutual solubility were chosen [87...89]. It was necessary to find out how the mutual solubility of the source metals affects the shape of the interface during explosive welding. There is a whole range of questions concerning the formation of copper–titanium welded joints:

- whether intermetallic welded joints are formed and under what conditions;
- what structure does the the local melting zones have;
- whether splashes form in the presence of intermetallic welded joints;
- how do intermetallics affect the self-organization of the cusps during the formation of waves;
- whether transition states are formed and under what modes of welding;
- are the transition states different from those observed for copper – tantalum welded joints;
- how dangerous is melting in the presence of intermetallic welded joints.

Here, for the first time in the monograph, the results of the study of copper–titanium welded joint are presented. Previously, they were not cited in either Chapter 3 or Chapter 7, since they were obtained later. A comparison is made with the results of the study of copper–tantalum welded joints, and we are limited to only a few typical structures.

8.1. Material and research methods

Explosive welding was carried out by the Volgograd State Technical University, OJSC Ural Chemical Engineering Plant (Yekaterinburg). An explosive charge was placed on the upper (cladding) plate. Parameters of used welding modes are shown in Fig. 113. Both double-layer copper–titanium composites (mode 3) and three-layer copper–titanium–copper composites (the other modes) were obtained. Three-layer composites were obtained as follows: first, copper—titanium was welded, then the two-layer composite was welded with copper. The phase diagram for the copper–titanium system is shown in Fig. 114 [90].

8.2. Experimental results (copper–titanium)

8.2.1. Welded joints (4'), (4)

The three-layer composite contains two welded joints: (4') and (4) titanium–copper–titanium. They were obtained at the same velocity

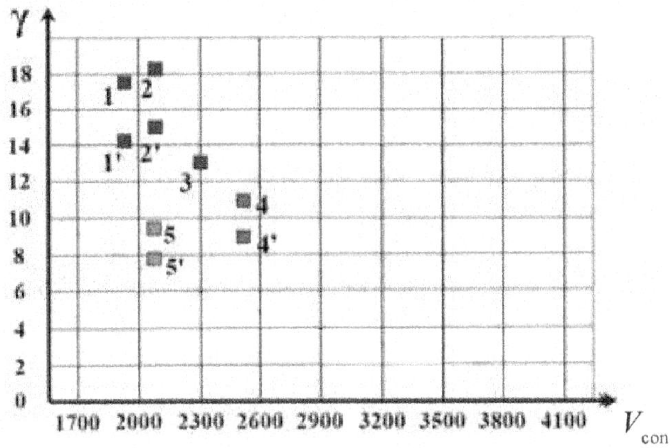

Fig. 113. Copper–titanium welds: welding parameters.

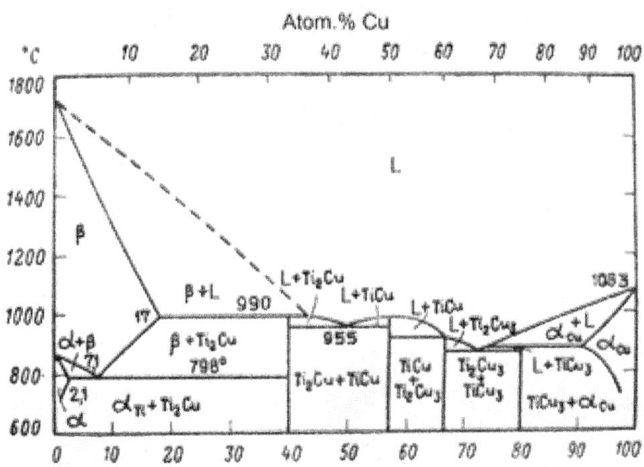

Fig. 114. Copper–titanium phase diagram.

of the contact point V_{con}, but with a different value of the angle of collision γ, which is approximately 2° less for the welded joint (4′).

Then, using the SEM (Fig. 115), the structure of the welded joints (4′) and (4) is investigated. As can be seen from Fig. 115 *a*, both boundaries of copper–titanium welded joints are wavy, but with different parameters. For welded joint (4′), the amplitude and wavelength are approximately 75 μm and 310 μm, and for compound (4), respectively, 150 μm and 610 μm. As usual, for a wavy boundary in the longitudinal section there are parallel bands. A comparison of

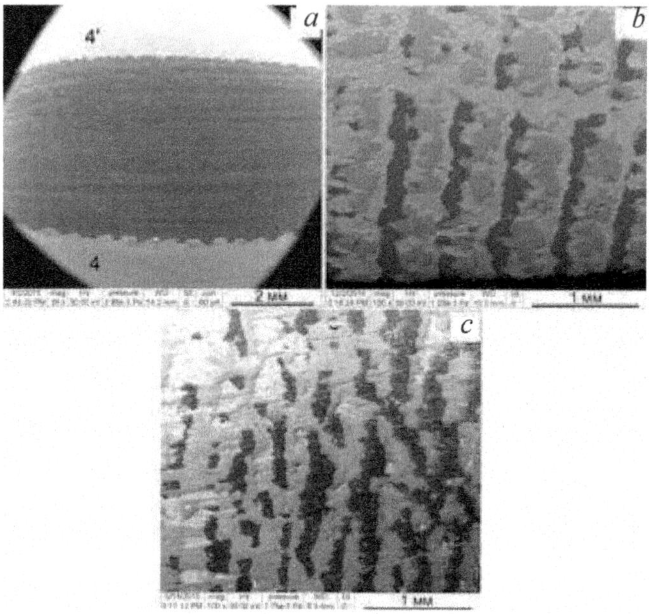

Fig. 115. Welded joints (4), (4'): *a* is the cross section for welded joints (4) and (4'); *b* – longitudinal section for welded joint (4); *c* – longitudinal section for welded joint (4').

Figures 115 *b, c* shows that the set of bands is less perfect for the welded joint (4'). At the same time, there are dark bands of titanium, two-coloured bands of copper, which contain unmelted copper, as well as lighter areas of local copper melting (then frozen). At the end of the chapter, the question of the structure of zones is discussed in more detail: either a copper–titanium solid solution or copper containing intermetallic particles.

As can be seen from Figs. 116, 117, waves on the surface of titanium for the copper–titanium welded joints are formed from cusps just as in the copper–tantalum welded joints studied in Chapter 7. However, the presence of mutual solubility of the starting materials has a dramatic effect on the structure of the waves. For copper-tantalum welded joints, it was shown that as the explosive effect intensifies, cusps first appear, then cusps and waves, then only waves. In contrast, for the copper–titanium welded joints, the cusps do not disappear. For welded joints (4'), this is clearly seen in Fig. 116, *a, b*. It is also seen that the cusps are strongly cut. In mode 4', a transition state is observed, which can be called a discontinuous wave (Fig. 116 *c*). The cusps form a wave at a certain length, then

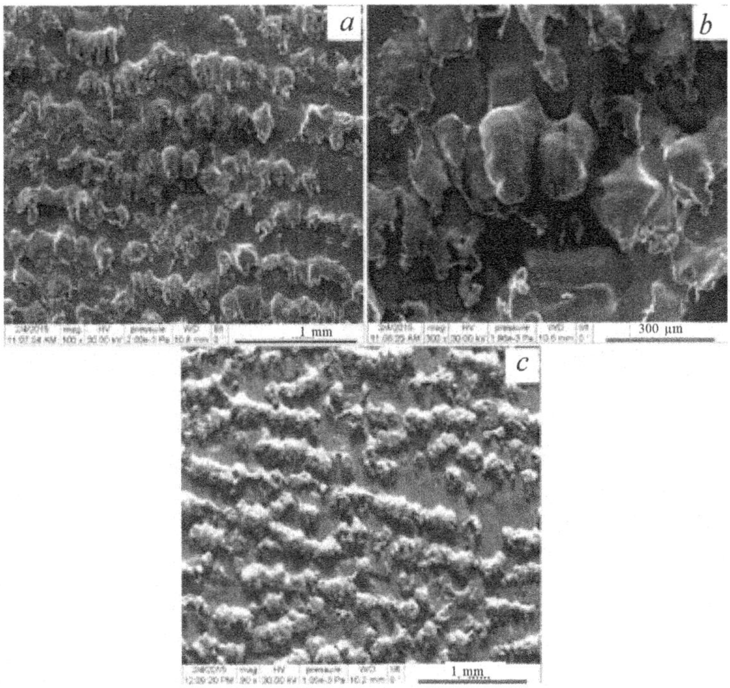

Fig. 116. Welded joints (4'): *a, b* – cusps, forming waves; *c* – intermittent waves.

the wave breaks, then again the cusps appear, etc. In this case, the direction of the wave axis is not very strictly maintained. Sometimes isolated cusps are visible.

The intermittent nature of the waves is also observed under mode 4 (Fig. 117 *a, b*), stronger than 4′. In this case, there is a more perfect wavy surface, but also containing, although to a lesser extent, wave breaks.

We believe that during explosive welding, self-organization of the cusps occurs, leading to the formation of waves. The processes of self-organization occur differently depending on the presence (absence) of mutual solubility of the starting metals. This is due to the fact that the mutual solubility makes diffusion and intermetallic reactions possible. Particles of intermetallics are visible in Fig. 117 *a, c*. It is the presence of intermetallic particles that prevents the mutual absorption of the cusps and the formation of waves free of cusps.

8.2.2. Welded joints (3)

For a two-layer composite (3), SEM images of the transverse and

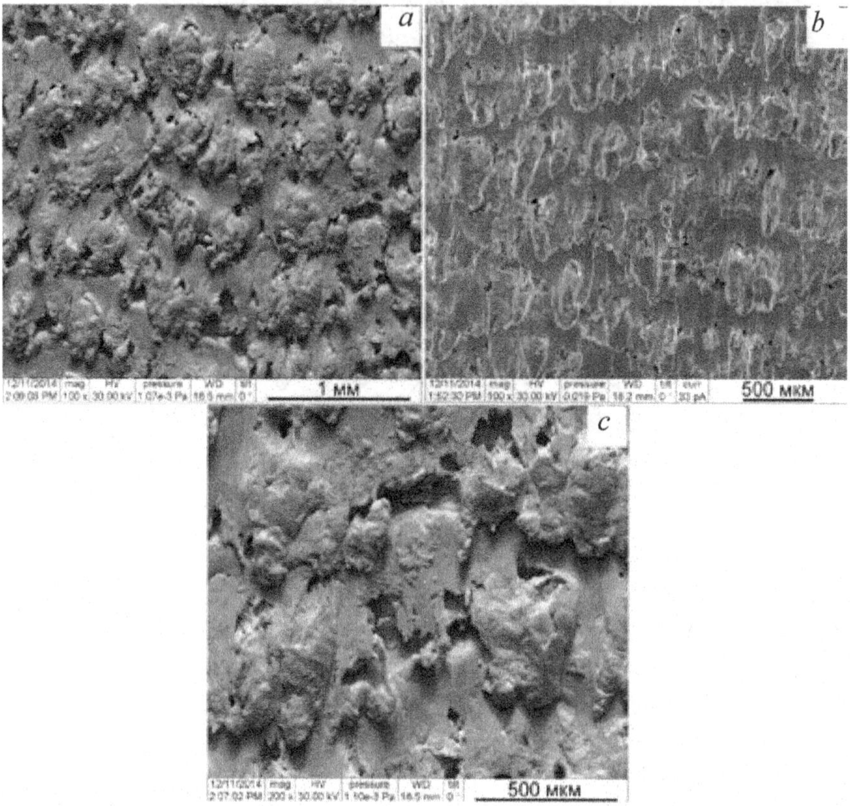

Fig. 117. Joint (4): *a, b* – intermittent wave (at different magnification); *c* – intermetallic particles on the surface of the cusps.

longitudinal sections are shown in Fig. 118 *a, b*. The wave-like boundary is more developed than for welded joints (4′) and (4). The amplitude and wavelength for welded joint (3) are approximately 150 μm and 450 μm, respectively. On the longitudinal section there is a set of parallel bands. But even in this case, as seen in Fig. 118 *b*, some bands either break off or change the direction of their axis in a certain section. Here, as in Fig. 115 *b* for welded joint (4), there are dark bands of titanium, two-colour bands of copper, which contains both unmelted and molten (then solidified) copper. For welded joint (3), the dimensions of the local melting zones are much smaller than for welded joint (4).

In the case when copper is etched, the surface of titanium looks like that shown in Fig. 118 *c*. This is a quasi-wave surface, for which all waves have approximately the same direction of the axis, but the waves are collected in long strips ('patches'). They are separated by narrower bands, where waves with a different direction of the axis are visible, or there are no waves at all. Such a quasi-wave surface can be considered as a combined transition state, namely a "patchwork quilt" consisting of intermittent waves combined into stripes. In fact, two transition states are observed in copper–titanium welded joints: a discontinuous wave (compound (4') – Fig. 116, *c*, welded joint (4) – Fig. 117 *a*) and a patchwork quilt (welded joint (3) – Fig. 118 *c*). A transitional state of another type with a chaotic distribution of 'patches', similar to that observed in Cu–Ta, was not found for Cu–Ti welded joints.

In Fig. 118 *d, e*, the cusps are tightly pressed to each other, the tops of which are cut up. The surface of the cusps is covered with small particles, which, as shown below, are intermetallic compounds. It is because of these particles that the coagulation of the cusps is difficult.

Images of various types of this patchwork on Fig. 119 imitate different forms of the quasi-wave surface arising from explosion welding.

8.2.3. Welded joints (1) and (1')

For a three-layer composite containing welded joints (1) and (4–1') the cross section of both interfaces is shown in Fig. 120 *a*. It is seen that for welded joint (1) the interface is wavy with an amplitude of approximately 45 μm and a wavelength of 190 μm, but the periodicity of the structure is disturbed, the amplitudes of the waves change. For the welded joint (1'), obtained with less exposure, the boundary is practically flat. Figure 120 *b* shows the SEM image of the transition state of the titanium surface for the welded joint (1) (copper etched). Such a transition state, like that observed for welded joint (3) (Fig. 118 *c*), contains alternating stripes resembling a patchwork quilt. Narrow lanes are clearly visible where the failure of the periodic structure occurs.

Figures 120 *c, d* is an SEM image of the titanium surface for the welded joint (1') (copper etched). It can be seen that here, in contrast to welded joint (1), there are not waves, but splashes. In accordance with the results presented in Chapter 7 for copper–tantalum welded joints, it can be assumed that the welded joint (1') was obtained near the lower weldability limit.

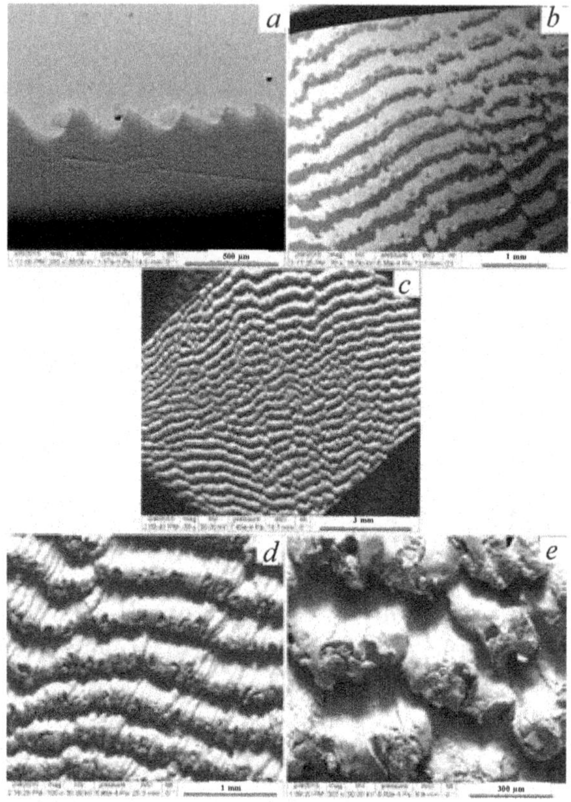

Fig. 118. Welded joints (3): *a, b* – wavy surface (transverse and longitudinal section); *c* - the structure of the type of 'quilt'; *c, d* – dense packing of the cusps at different magnifications (*c, d, e* - copper etched).

8.2.4. Welded joints (2) and (2')

The cross-section for both welded joints is shown in Fig. 121 *a*. It is seen that for welded joint (2) the interface is wavy with an amplitude of approximately 40 μm and a wavelength of 170 μm. For a welded joint (2'), the boundary is almost flat. In Fig. 121 *b* a longitudinal section is given for welded joint (2), but the section plane is inclined with respect to the interface. Alternating bands of copper and titanium can be seen. In the case when copper is etched, the titanium surface has the form shown in Fig. 121 *c*. There are no intermittent waves, no quilt-type structures. But the wavy surface is still not perfect, since the axes of the waves change. For the welded joint (2'), the titanium surface structure (Fig. 122 *a*) is close to the observed one (Fig. 120 *c*) for the welded joint (1'). The surface of titanium is covered with splashes (Fig. 122 *b, c*) in the same way

Fig. 119. Different types of 'quilt'.

as in the case of a welded joint (1′) (Fig. 120 *d*).

The similarity of the structures observed for welded joints (1') and (2') is clearly visible. This means that both of these welded joints are obtained near the lower boundary (LB) of weldability. At the same time, welded joints (1) and (2) are obtained above (LB). At the same time, the welded joint (2), for which there is a fairly perfect wavy interface (Fig. 121 *c*), is obtained at a greater distance from LG than the welded joint (1), for which a quilt-type surface is observed (Fig. 120 *b*).

8.2.5. Welded joints (5) and (5′)

For the copper–titanium–copper three-layer composite, the resulting modes (5) and (5'), a fairly smooth surface area (Fig. 123 *a*) corresponds to the mode (5'). At first, it seems that welding has occurred, but during subsequent polishing the sample disintegrates. This means that for the mode (5') the parameters γ, V_{con} are below

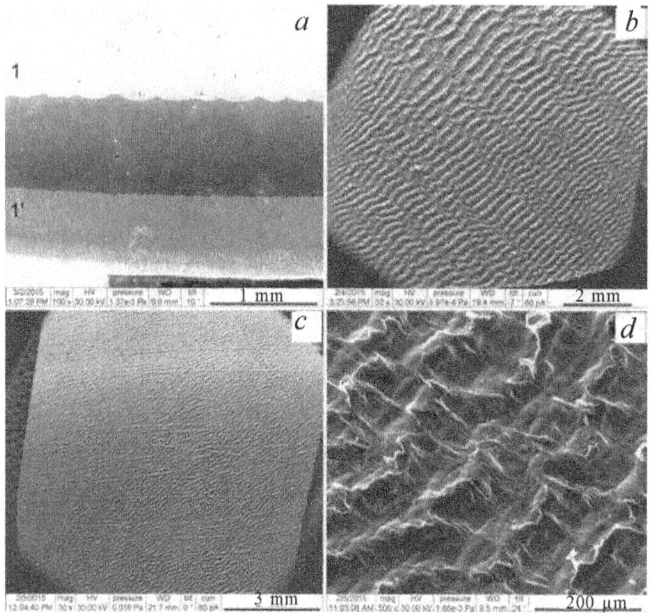

Fig. 120. Welded joints (1) and (1'): *a* – cross section for two interfaces; *b* – quilt structure (welded joint (1); *c* – titanium flat surface (welded joint (1')); *d* – splashes (welded joint (1')).

the LB.

Mode (5) with its parameters, on the contrary, falls into the region slightly higher than LB, although near it. In this case, the surface of titanium has the form shown in Fig. 123 *b*...*d* (copper etched). One can see dense groups of cusps pressed to each other, which are lined up in rows, but do not form waves (Fig. 123 *b*). As can be seen from Fig. 123 *c*, the surface of titanium, containing similar groups of cusps, is heterogeneous. In some places, it contains separate misoriented areas ('patches'), becoming like a 'patchwork quilt', albeit without waves (Fig. 123 *d*).

As follows from the foregoing, the curve $\gamma - V_{con}$, defining the lower border of the weldability window, will pass in Fig. 113 between points corresponding to the modes (5) and (5'). It can be assumed that with a decrease in the velocity V_{con}, the indicated curve goes steeply upwards, since splashes are observed under the modes (1') and (2'), as mentioned above. The proposed LB move is as follows: between points (5) and (5'), then near the point (1'), but slightly to the left of it, i.e. with a smaller value of the parameter V_{con}, then to the left of the point (1).

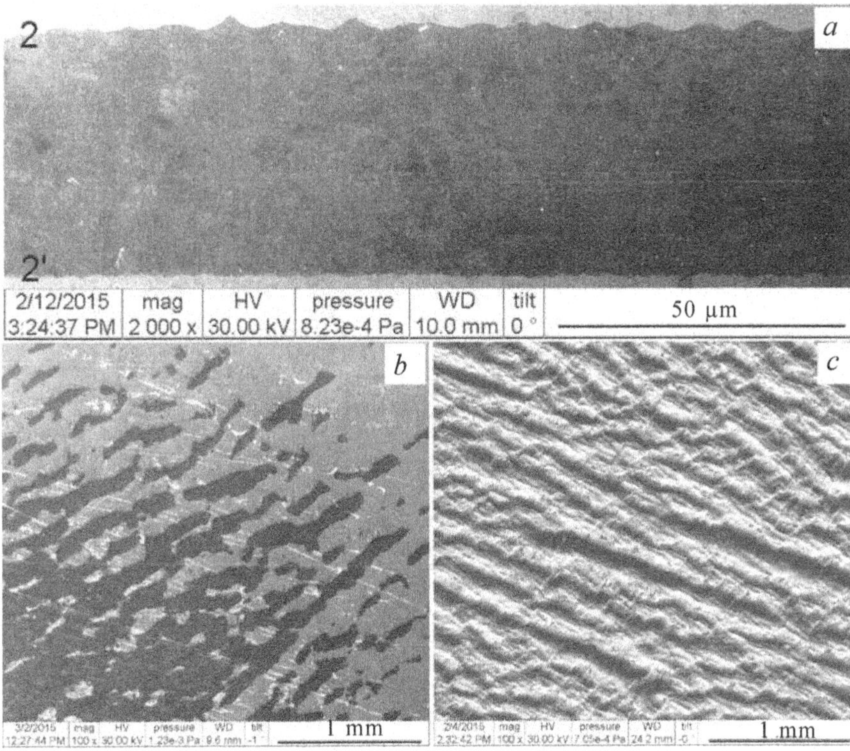

Fig. 121. Welded joints (2) and (2′): *a* – cross section for two interfaces; *b* - longitudinal section for the welded joint (2); *c* – a wavy surface of titanium (copper etched), welded joint (2).

8.2.6. The formation of intermetallic welded joints

Heterogeneities of the interface were observed in all welded joints investigated in the monograph: cusps and local melting zones (after solidification). A typical three-colour picture of the longitudinal section of the transition zone for a flat interface is shown in Fig. 30. For each of the welded joints ($\mathbf{C}_p$) copper–tantalum and ($\mathbf{E}_p$) aluminum–tantalum, differing in the type of mutual solubility of the original elements, three areas are visible: tantalum cusps, copper (aluminum) areas that did not melt, and local copper melting zones (aluminum) containing particles of a different phase.

As was shown above, in the case of ($\mathbf{C}_p$) welded joints these are tantalum particles, and in the case of the ($\mathbf{E}_p$) welded joint they are intermetallic Fig. 30. Figure 42 *b* shows the scattering of such particles, and Fig. 42 *c* shows that the particles are poured out of

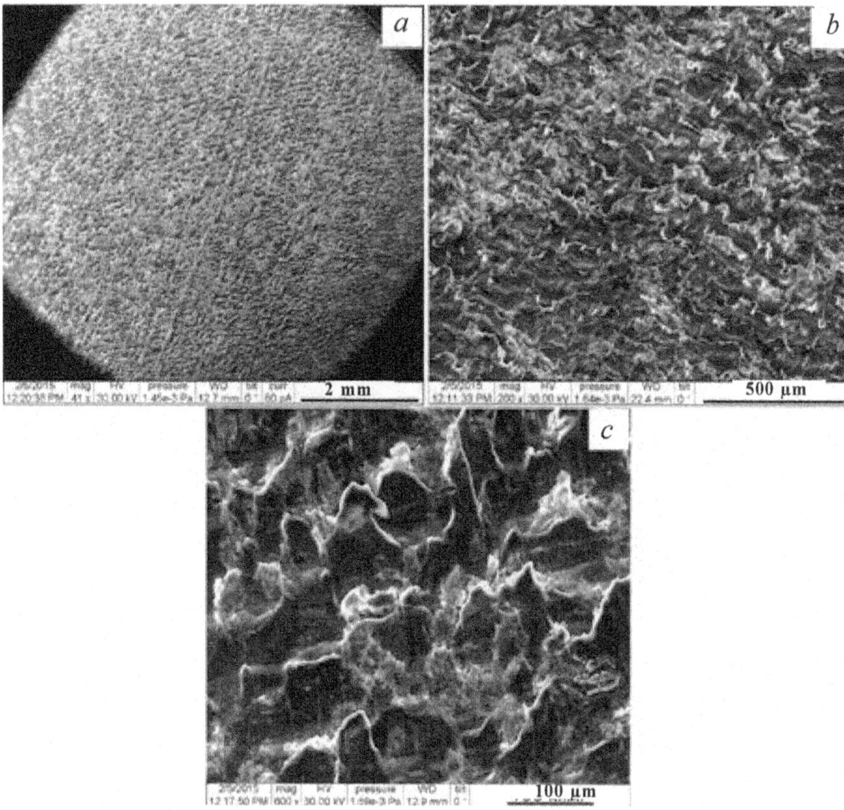

Fig. 122. Joint (2'), longitudinal section (copper etched): *a* – flat surface of titanium; *b, c* – splashes at different magnification.

the zones of local melting during the etching of aluminium (welded joint (**E**$_p$)). Similarly, as seen in Fig. 31 *b*, tantalum particles are poured from local melting zones during copper etching (welded joint (**C**$_p$)). In fact, Fig. 42 *c* and Fig. 31 *b* show the main difference in the structure of local melting zones, due to the presence (or absence) of mutual solubility. The TEM image of the microheterogeneous structure of the local melting zone for the welded joint (**C**$_p$) is shown in Fig. 33. This zone is filled with tantalum nanoparticles inside the copper matrix.

For a wave-shaped interface, as noted many times, the longitudinal section is a set of bands. As can be seen from Fig. 37 *a*, for the (**C**$_w$) welded joint, cusps and local melting zones (rounded areas of white colour) are visible on the boundaries of the bands. A swirling of the vortex type is observed inside the zone (Fig. 37 *b*). In this

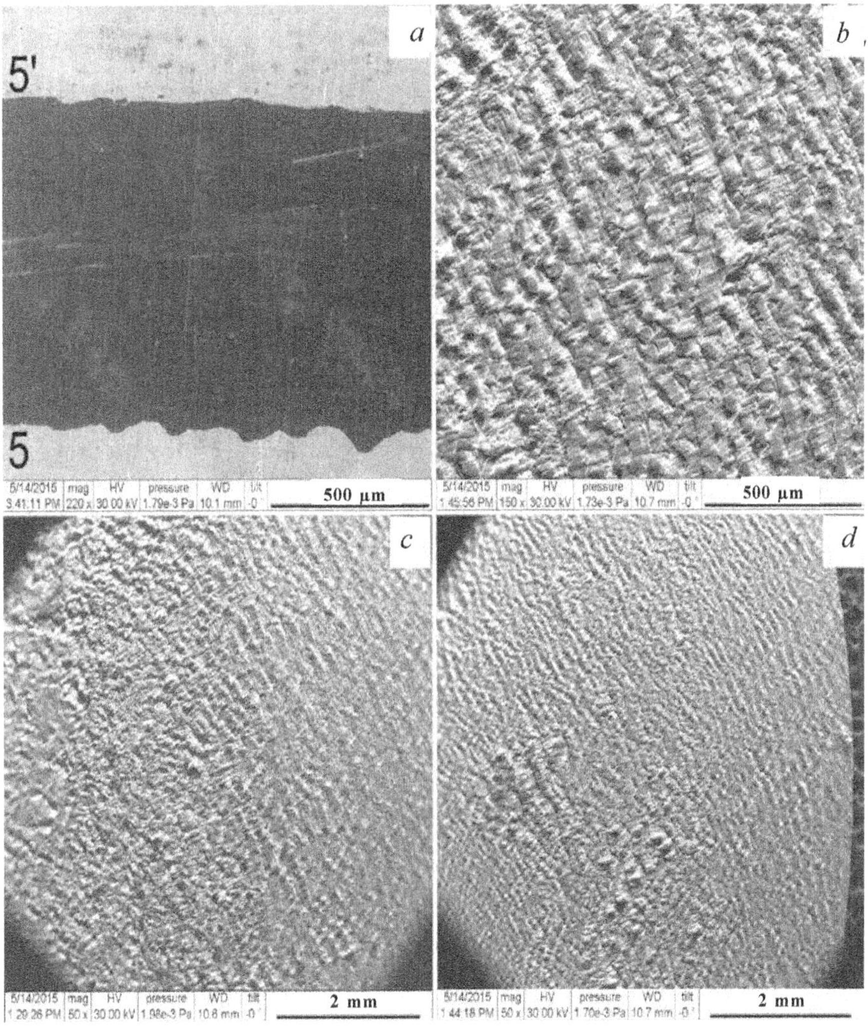

Fig. 123. Welded joints (5') and (5): *a* – cross section for two interfaces; welded joint (5), copper etched, *b* – cusps; *c, d* – heterogeneous surface of titanium containing cusps.

case, the vortices are imperfect compared with those observed for titanium–orthorhombic titanium aluminide welded joints (Fig. 13). The microheterogeneous structure of the local melting zone is shown in Fig. 38 *b*. Such a structure is similar to that shown in Fig. 33 for the ($\mathbf{C}_p$) welded joint, and contains, in contrast, micron tantalum particles.

For the aluminum–tantalum welded joint ($\mathbf{E}_w$), which has a wave-like boundary, no local melting zone was observed. Instead, a near-surface layer was observed in the form of a film covering a wave-like surface. The melted (then frozen) film (Fig. 46) contains particles of a different phase, the dimensions of which vary widely: from 50 microheterogeneous structure of the local melting zones for Cu–Ti welded joints was studied in [88, 91–93].

For welded joint (2) Fig. 124 shows a longitudinal section of the interface. From Fig. 124 *a* we can conclude that this surface is quasi-wave. With a higher magnification (Fig. 124 *b*), a zone of local melting was found. A similar conclusion can be drawn from the chemical analysis of this zone (Fig. 124 *d*). With a larger increase in Fig. 124, particles are visible inside the molten (and then frozen) copper. Some of them are smooth, others are faceted. It can be assumed that these particles are respectively either clusters or intermetallic compounds.

Local melting zones for copper–titanium welded joints (4) are presented in Fig. 115 *b*. With a larger magnification than in Fig. 115 *b* (approximately twenty times), the vortex structure of the zones was found (Fig. 125 *a*). The vortices are imperfect, as in the case of the copper–tantalum ($\mathbf{C}_w$) welded joint considered above (Fig. 37 *b*).

But in any of these cases, the vortices are imperfect compared with those observed for titanium–orthorhombic titanium aluminide welded joints [60, 61]. Inside the local melting zone, outside the vortexes, in Fig. 125 *b* a region containing numerous particles is visible. This area at a higher magnification has the form shown in Fig. 125 *c*. The zone is filled with molten (then frozen) copper, which, unlike the copper–tantalum welded joints, contains particles of intermetallic welded joints. The corresponding diffractogram for welded joint (4) is shown in Fig. 125 *d*. The observed lines belong to the Cu_3Ti intermetallic compound.

A TEM image of the structure of the local melting zone (welded joint (4)) is shown in Fig. 126. As a result of TEM analysis, the following phases were identified: the metastable phase Cu_3Ti (Fig.126 *a–c*) and the equilibrium phase Cu_4Ti (Fig.126, *d–f*), denoted as β-phase in Fig. 114. The metastable phase Cu_3Ti has an orthorhombic lattice ($D0_a$). The equilibrium phase Cu_4Ti has the structure Au_4Zr. As follows from the comparison of Figs. 125 and 126, X-ray diffraction analysis makes it possible to identify only the Cu_3Ti phase, while TEM analysis revealed also particles of the Cu_4Ti phase in some areas of the local melting zones.

The microheterogeneous structure of local melting zones for Cu – Ti welded joints (mode (3)) was investigated after explosive welding and subsequent heating [92, 93]. Figure 118 shows a wavy interface (cross-section and longitudinal section), as well as the structure of a patchwork-type interface surface (copper etched).

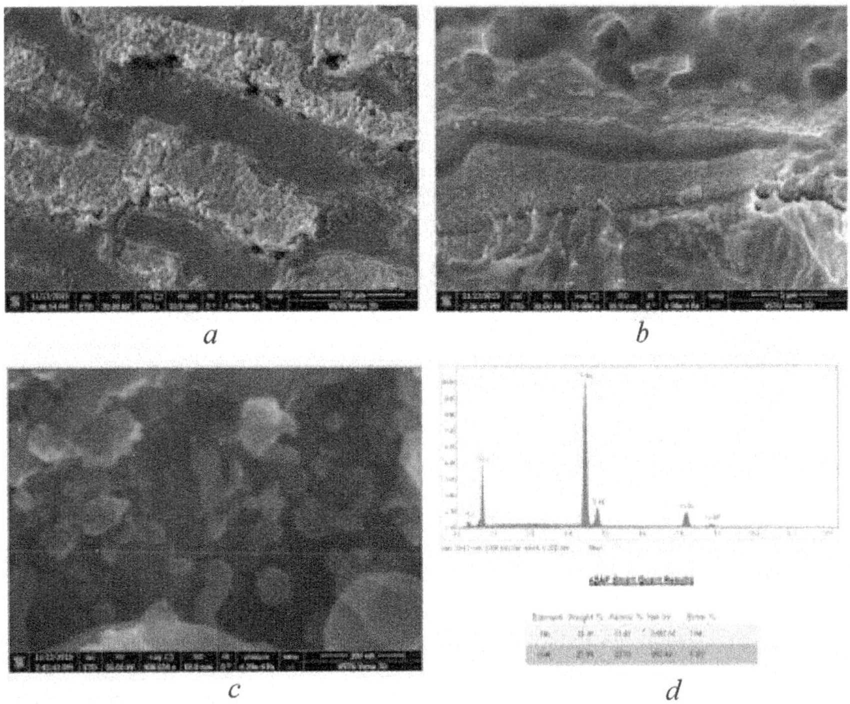

Fig. 124. Copper–titanium welded joint (2), quasi-wavy interface, longitudinal section: *a* – copper and titanium bands, *b* – melt zone, *c* – intermetallic welded joints in the melt zone, *d* – chemical analysis of the melt zone.

In Fig. 127, and the SEM image of the longitudinal section of the wave-like border is shown with a larger magnification compared to that at which Fig. 118 *b* was produced. Alternating copper and titanium bands, cusps and local melting zones at the boundaries of the bands are visible.

In Fig. 127 *b*, the vortex structure of the section (in the frame) of the indicated zone is visible. With a larger magnification, numerous luminous particles are observed against the background of the vortex structure (Fig. 127 *c*). As can be seen from Fig. 127 *d*, most particles are smooth. As mentioned above, these are apparently particles of clusters. Also visible are faceted particles of intermetallic compounds.

Using the attachment to the scanning microscope, the sample was heated and the video recording of the change in its structure. Several frames are shown in Fig. 128. It can be seen how the initially rounded

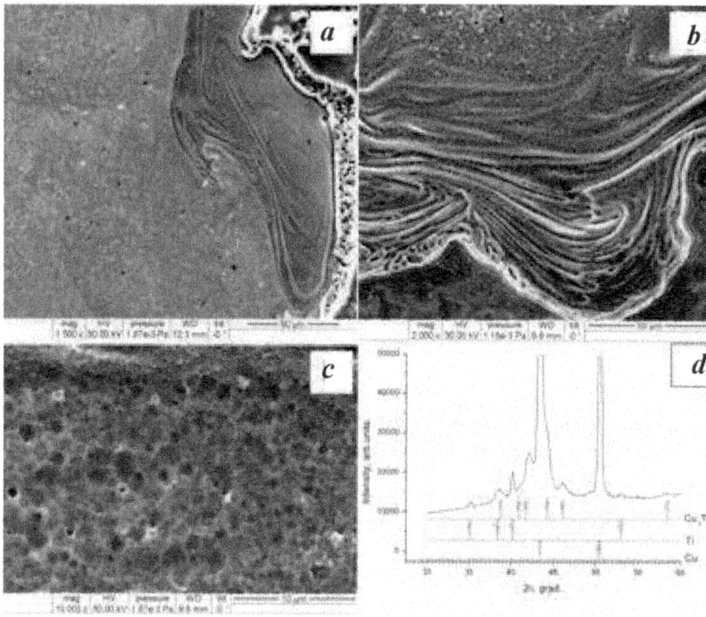

Fig. 125. The internal structure of the local melting zone (compound (4)): *a* – vortices; *b* – vortices and particles; *c* – particles of clusters and intermetallic welded joints outside the vortices; *d* – diffraction pattern of the local melting zone).

particle, as the heating time increases (indicated in each shot), acquires a certain face, becomes elongated and unlike the original one. The appearance of the cut is evidence of the transformation of clusters into intermetallic welded joints. Moreover, the convergence of particles is visible, leading to the absorption of small particles by large ones.

To identify the phase composition of particles found at the interface of copper–titanium welded joints, X-ray analysis was performed. The surveys were made from the surfaces of thin sections of longitudinal section samples obtained by removing copper layers with abrasive paper until traces of titanium appear. This is in contrast to the filming of a sample of the Cu–Ti welded joint produced by explosive welding using mode 4, the diffraction pattern of which was taken from the cross-section of the thin section and is shown in Fig. 125 *d*. The X-ray diffraction pattern of the Cu–Ti welded joint obtained by explosive welding according to the mode (3) reliably records X-ray peaks of the following phases: titanium, copper, Cu_3Ti (Fig. 129 *a*). However, when fitting the profiles of diffraction reflections, the best result in matching the calculated and

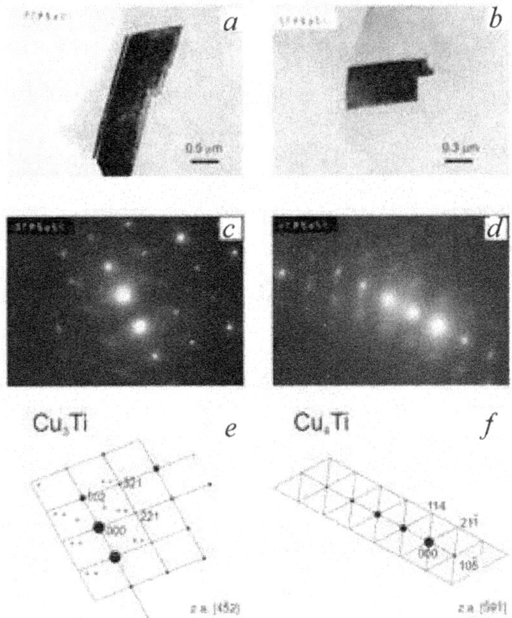

Fig. 126. TEM image of the local melting zone of welded joint (4): *a* is a particle of the Cu_3Ti phase, *b* is microdiffraction for (*a*); *c* – decoding of microdiffraction; *d* is a particle of the Cu_4Ti phase; *e* – microdiffraction for (*d*); *f* – decoding of microdiffraction (z.a. is the axis of the zone).

experimental diffraction patterns was achieved by adding peaks from the intermetallic phase CuTi to the calculated diffraction pattern, as well as the peak (204) of the Cu_4Ti phase.

Figure 129 *b* shows the diffraction pattern from a sample of the Cu–Ti welded joint obtained by explosive welding according to regime 3 and subjected to annealing at a temperature of 500°C for 1 hour. Unlike the diffraction pattern of the sample without heating (Fig. 129 *a*), this diffraction does not contain peaks of the intermetallic phase Cu_3Ti. However, in addition to the already observed peaks of the CuTi phase, in this sample one can reliably identify not one but several peaks of the Cu_4Ti phase. As in the previous case, due to the best fit of the calculated and experimental diffractograms, fitting the diffraction reflection profiles reveals traces of another intermetallic phase, namely the Ti_2Cu_3 phase.

Thus, as a result of hourly heating of the sample at 500°C, we observe the disappearance of the X-ray peaks of the Cu_3Ti phase and the appearance of peaks from the Cu_4Ti and Ti_2Cu_3 phases. If

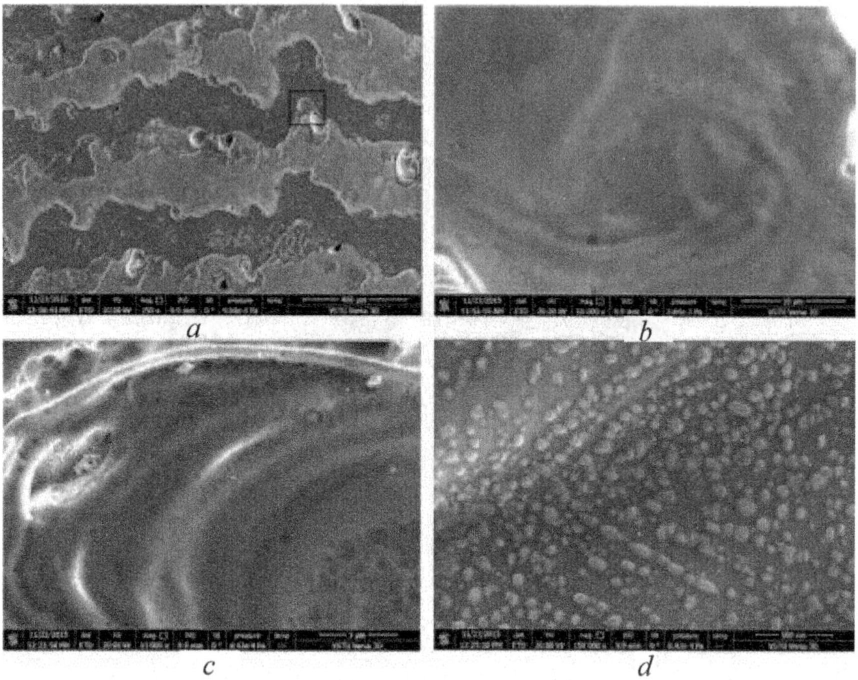

Fig. 127. Microheterogeneous structure of local melting zones for the compound (3): *a* – longitudinal section of the wave-like surface; *b, c* – vortex structure of the zone with different magnification; *d* – particles of clusters and intermetallic welded joints inside the zone.

we take into account that, according to [94], the Cu_3Ti phase is metastable and decomposes to Cu_4Ti and Ti_2Cu_3 when heated to 500°C, then the result is quite natural. The presence of a single peak of the Cu_4Ti phase on the diffraction pattern of the sample without heating is explained by the initial stages of the decomposition of the metastable Cu_3Ti phase, which according to [95] can occur even at room temperature. The heating up to 500°C used in the work only accelerated this decay.

In conclusion, we note that, according to generally accepted concepts, the formation of intermetallic welded joints during explosion welding can be dangerous for the strength of a joint. New particles, as a rule, are not born during heating, but existing particles grow due to the processes of environmental decay. To some extent, this process is similar to Oswald ripening, which is a process of condensation of supersaturated phases in the later stages of development, when the nucleation stage ends and the growth of large particles occurs due to small ones [96...98]. The name of this

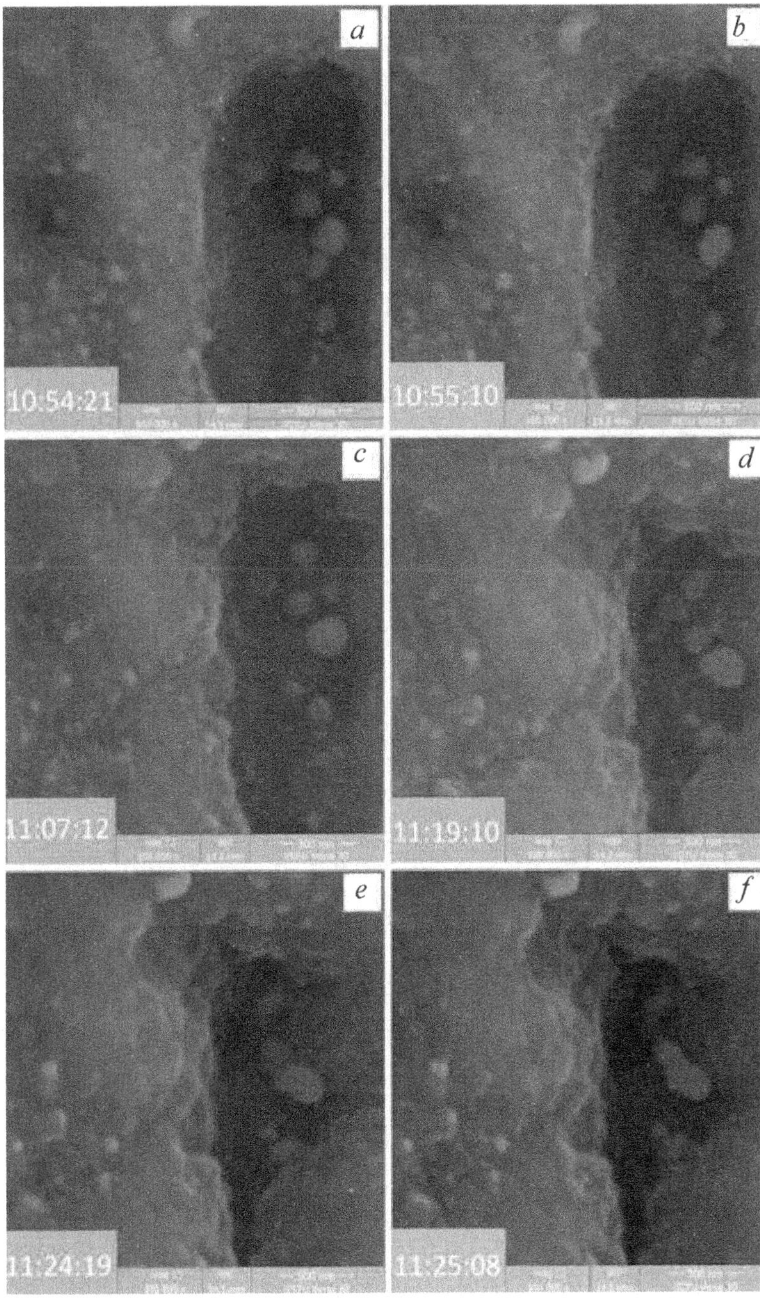

Fig. 128. Changes of the shape of a particle during heating.

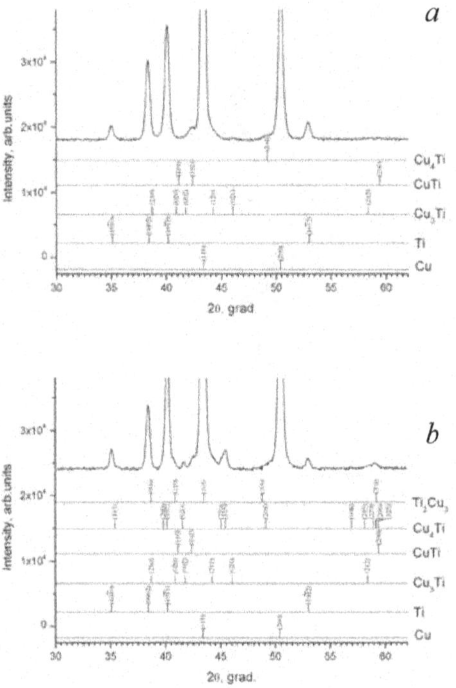

Fig. 129. The diffractogram of a sample of Cu-Ti welded joint: *a* – mode (3) without annealing, *b* – mode (3), annealing at 500°C for 1 hour.

process is used here in a collective sense and implies structural changes occurring after the formation of particles.

The solid intermetallic sample is brittle. On the other hand, hardened melt inserts with intermetallic particles are used to harden, for example, steels and other materials. Indeed, quite often there are very strong welded joints containing initially molten and then frozen regions, within which there are numerous individual intermetallic particles. We believe that such particles become dangerous only in certain cases, in particular, with the intensification of the welding mode. In this case, one can imagine the dynamics of an ensemble of intermetallic inclusions in the following form: first, the individual particles are enlarged, and then they are combined into some agglomerates. In this case, some particles absorb other, smaller ones. Such a process as applied to recrystallization was figuratively called 'cannibalism' by Kann [98].

The impetus for this is that instead of two surfaces, one appears. But due to disorientation of the crystal lattices, a mismatch may occur when they are joined, which leads to an increase in the surface energy. A similar process of sticking powder particles leads to its consolidation with such a strong impact as torsion under pressure.

When the critical size is reached, the agglomerates behave like a solid body. This possibility can be realized, for example, near the upper boundary of the weldability window. However, depending on the melting point of the starting materials, the temperature of formation of intermetallic compounds, the surface energy of the intermetallic–metal, the occurrence of continuous or almost continuous significant volumes of the intermetallic phase can occur under less severe conditions.

Since the process leading to the adhesion of particles is fairly general, for various welding conditions and different conditions of subsequent operation (load, heating) there is always the possibility of destruction of the joints due to the fragility of intermetallic phases. The variety of modes and operating conditions makes the danger of formation of intermetallic welded joints quite real.

Conclusions to chapter 8

1. The copper–titanium welded joints were found to contain cusps in the form of splashes on the flat surface of titanium, similar to those observed for copper–tantalum welded joints. In both cases, the cusps, although solid, are similar to splashes that occur on the surface of water.

2. An unusual waveform was detected: in the presence of mutual solubility wave is a dense packing of the projections. It can be assumed that this is due to the presence of intermetallic phases on the surface of the cusps. Such a structure can serve as direct evidence of the self-organization of the cusps as a method of forming waves.

3. It is shown that the transition from splashes to waves occurs through a series of intermediate states. One of them is a discontinuous wave. A combined transition state, which is formed by intermittent waves, combined into bands is also detected. Such a structure is similar to a strip-type patchwork 'quilt'.

4. It is shown that the local melting zones have a vortex structure and contain numerous particles of clusters and intermetallic welded joints.

5. When heated after explosive welding, the shape of the particles changed: a faceting occurs on smooth particles, which is evidence of the transformation of clusters into intermetallics.

6. X-ray diffraction analysis showed that the Cu_3Ti intermetallic phase observed after welding is thermally unstable and turns into other intermetallic phases upon heating.

9

Welding of homogeneous materials

The main requirement for the mechanical properties of composites determining their reliability and durability is, first of all, a guaranteed assurance of continuity and high strength of the bonding of layers. The extensive theoretical and experimental material accumulated to date, as well as the created physical and mathematical models of the process under consideration, allow us to purposefully control the properties of welded joints, including precisely dosed energy saving in the impact zone for a particular pair of materials being joined.

9.1. The structure and properties of explosion-produced joints of homogeneous metals and alloys

9.1.1. Bimetals from aluminium and its alloys

The properties of Al-based alloys vary in a wide range due to alloying, as well as strain and thermal hardening. As noted in [99], technically pure aluminium has the widest range of weldability. It should also be noted that strong joints are produced in a fairly wide range of contact point speeds of 1200–1500 m/s. In the group of medium strength alloys, hardened by deformation or heat treatment, aluminium alloy 1201T1 has a relatively wide range of weldability.

High-strength aluminium alloys have the narrowest range of weldability, and according to [99], it is not possible to achieve the equal strength welded joint V95T1 + V95T1. Starting from a certain energy, the joints cannot be maintained: the plates are peeled apart from each other, although wave formation characteristic of explosive welding is observed on the surfaces of the separated plates. In this case, a fine porous mass of material is found in the troughs of the waves.

It is obvious that the range of weldability of aluminium alloys can be expanded in two ways: (1) by reducing the required level of critical energy input during welding by annealing heat-treated alloys with subsequent strengthening heat treatment; (2) an increase in the permissible level of marginal energy input.

9.1.2. Steel bimetals

Explosive welding of homogeneous and dissimilar steels, as a rule, does not cause serious difficulties, since their weldability is very high both in terms of energy input and in contact speeds.

In the joint zone of homogeneous low- and medium-carbon steels, heterogeneity of the structure and properties in the form of molten metal with a martensitic structure is observed. Heating of such welded joints leads to the disordering of the structure and the complete decomposition of martensite. The temperature of these processes depends on the composition of the steel and the degree of its hardening. The heterogeneity of the structure and properties in the joint zone of like low- and medium-alloyed steels after explosive welding can be reduced or completely eliminated by appropriate heat treatment [100].

Welded joints of the same high-alloyed austenitic chromium–nickel steels, obtained in a wide range of modes, have high strength, the value of which usually exceeds the strength of the raw materials. The heterogeneity of the structure in the joint zone in the form of melted and hardened metal sections increases the strength of the joint, but at the same time, reduces its plasticity. In this case, the hardening of the contacting layers of metals is higher, the lower the initial hardness of the material.

The holding of hardened welded joints of austenitic steels in the temperature range of 500-800°C leads to an intense precipitation of carbides along the grain boundaries and slip planes, i.e. in places with a high density of defects in the crystal structure. In areas of the cast phase, where the hardening is absent, the precipitation of carbides does not does not take place.

Thus, the treatment of welded joints in homogeneous austenitic steels in the temperature range of 500–800°C is undesirable, since the hardened layers of the WZ (weld zone) have a high tendency to decomposition, as a result of which their resistance to intercrystalline corrosion and impact viscosity are reduced compared to the untreated metal.

When explosive welding high-alloyed ferritic steels with carbon steels, the main problem is the precipitation of chromium and titanium carbides on the formed parts of the cast metal. In addition, the decarburization of carbon steel occurs in the processing of such compounds in the weld zone, and a layer of carbides is formed in the ferritic alloy steel.

The explosive welding of pearlitic and austenitic steels leads to their significant strengthening in the joint zone [101].

Considering the fact that the bimetal in question is overwhelmingly used in the manufacture of pressure vessels under alternating loads at high and cryogenic temperatures, it is of practical interest to study its resistance to brittle fracture and cyclic loads.

In [5], the positive role of the cladding (usually more ductile) austenitic layer of structural steel is noted from the point of view of the strain capacity and susceptibility to unstable fracture. At the same time, there is a need to take into account a high level of local heterogeneity, which determines the active deformation in the weak ferritic zone of the compositions. Its local foci are associated with the development of microcracks in the adjacent brittle areas. These cracks nucleate long before the destruction of the entire material. Plastic layers, on the other hand, cause inhibition of the free movement of microdamages. In this case, the final behaviour of the composition is determined by the predominant action of one of these opposing processes. With their proper ratio, the strength indicators can be increased, especially noticeable when cladding high-strength steels [5].

When testing the steel 08Cr18Ni10Ti + steel 22K bimetal for resistance to fatigue failure, it was found that the endurance limit of the bimetal achieves the corresponding characteristics of the base metal or somewhat exceeds them. In this case, fracture developed predominantly along the base metal, and in the bonding zone, up to the final stage of the experiment, no visually detected damage was observed. The metallographic analysis of the weld zone also did not reveal any damage at the microlevel either in the decarburized layer or in the adjacent solid interlayers.

When evaluating low-cycle behaviour of welded steel compositions in the stage of damage, according to [5], the best group included steel clad with austenitic chromium–nickel steels, subjected to high-temperature heat treatment and having a defect-free structure with an ordered wavy structure and minimal content of cast inclusions. For them, the average strain at which damage occurs in the weld zone is approximately 20%. The welded joints in the same combinations of materials, but having structural defects, had an average strain of about 4% at N = 10–20 cycles [5].

9.2. The choice of a homogeneous copper–copper pair

In previous chapters, we established patterns of change in the structure of the interface for welded copper–tantalu and copper-titanium pairs, which are realized under conditions of intensification of the welding mode. The differences in the mutual solubility of the coupled metals allowed to reveal the unique features of the formation of a quasi-wave relief. Along with this, it was found that the features of the evolution of the transition of structures during the intensification of the welding regime are practically independent of mutual solubility.

In order to expand the range of the studied compounds, as well as to determine whether the identified features are universal, an additional copper–copper metal pair was analyzed [102–105]. The choice of this homogeneous pair is not accidental: for it a number of self-organizing mechanisms 'realized to nothing', which are obtained for copper–tantalum and copper–titanium. These include processes caused by both differences in the physical properties of the samples to be welded (difference in melting point, density, etc.) and the features associated with mutual solubility. Thus, the physical processes that occur during explosive welding for a homogeneous pair become smaller, and it follows from this that the remaining mechanisms become more defined. Copper was chosen as the starting material for the convenience of subsequent comparison with Cu–Ta, Cu–Ti welded joints.

However, to study the structure of the interface, it is necessary to be able to completely etch one material and observe the surface of another. This is precisely the difficulty of studying the interface for joining homogeneous materials. Therefore, a copper–copper pair is modelled by a copper–melchior pair. This choice is due to the fact that copper and nickel silver have similar properties: close melting point and density. This allows one to approximately consider them homogeneous. However, due to the fact that copper and melchior react differently to etching, have a different color and have a different chemical composition, the transition zone between these two materials is certainly noticeable during electron microscopic observation.

Summarizing the above, we note that the main tasks in the study of this welded pair were as follows:

•to investigate the features of the evolution of the interface with the intensification of the welding regime in the case of homogeneous metals;

•to identify the fundamental patterns associated with explosive welding, in the absence of specific features arising in a non-uniform welded pair;

•try to find transients between flat and wavy boundaries, similar to those found for copper–tantalum, copper–titanium welded joints;

•to determine what features the local melting zones have in the case of a homogeneous pair of materials being welded;

•to reveal the features of the fractal dimension for the copper-copper joint.

9.3. Welding parameters

A parallel arrangement of the plates was used: the top plate (copper M1) was set in motion by a controlled explosion, as a result of which it hit the bottom plate at high speed (melchior alloy MN19). We used nickel silver of the MN19 brand, which has a chemical composition: Ni – 18–20%, Cu – 78–80%, and with an impurity concentration 1.5%. The density of copper was 8.93 g/cm^3, and the density of the melchior 8.9 g/cm^3. Obviously, these values are close, so the conditions of the problem are met. In addition, melchior MN19 have similar melting points (T_{melt} = (Cu) 1085°C, T_{melt} = (MN19) 1190°C). However, a distinctive feature of melchior alloy (MN19) can be considered that in the annealed condition it has a greater tensile strength (290 MPa) compared with copper M1 (245 MPa) after annealing. Nevertheless, we can approximately consider the copper– melchior homogeneous.

The welding modes for the melchior alloy were conventionally divided into two groups. In the designation of the first numbers are used only: (1), (2), (3). The second group is denoted by the letters: (A), (B), (C), (D). Within the first group, a strict intensification of the welding mode from the lowest energy (3) to the highest energy (1) was chosen. In this case, only one parameter was varied: the contact point velocity (V_{con}). The second group is characterized by changes in both external parameters of the system (the contact point velocity V_{con} and the angle of impact γ). The intensity of the modes increases from (D) to (A). All investigated modes for the copper-melchior joint are presented in Fig. 130.

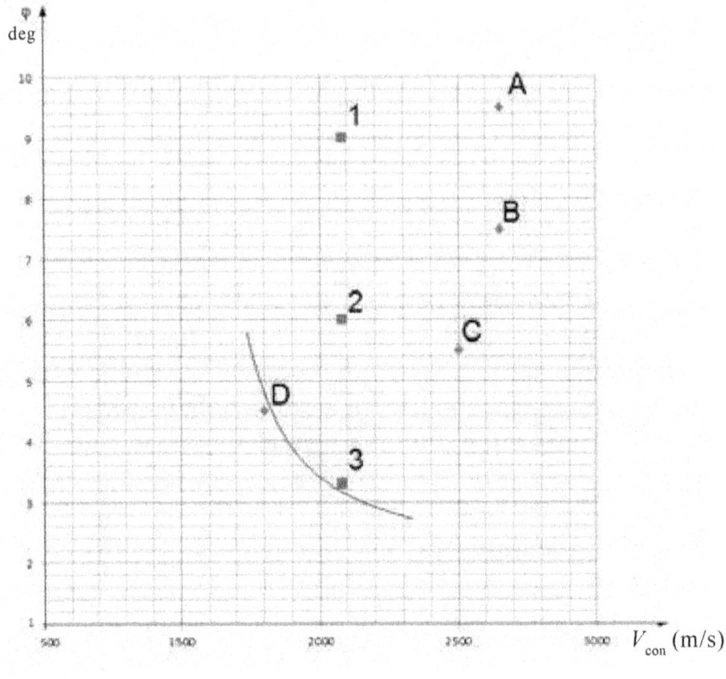

Fig. 130. Copper–melchior welded joint: parameters of welded samples (γ is the angle of impact, V_{con} is the velocity of the contact point).

Note that the differences between the elements of different groups are not so obvious: for example, it is impossible to say exactly which of the modes (2) or (**C**) has a high pumping energy, based on their location in the weldability window. A similar analysis will be carried out in the following sections, based on the microstructure of each welded joint, as well as the fractal dimension of the slices of these samples.

When performing research, longitudinal and cross-sections of the samples were used. It should be noted that the method of etching one sample so that the relief of the other is maintained, in this case, it does not work efficiently enough, since copper and melchior alloy are too close in chemical composition.

The parameters for welded specimens of copper–melchior compounds are listed in Table 11:

Table 11. Modes of explosive welding for melchior alloys

Mode	Impact angle γ, °	Contact point velocity V_{con}, m/s
(AB)	9.5°	2650
(B)	7.5°	2650
(C)	5.5°	2500
(D)	4.5°	1800
(1)	9°	2080
(2)	6°	2080
(3)	3.3°	2080

9.4. Experimental results for copper–melchior alloys welded joints

In order to find out what type of boundary is characterized by each of the above modes, the period and amplitude of the wave were determined (in cases where this can be done). Table 12 shows the values of these parameters depending on the welding mode:

Based on the amplitudes given above, a preliminary conclusion can be made about the energy characteristics for each of the modes. Starting from the highest energy mode – (A), and ending with the lowest energy mode – (2), the distribution is as follows: (A), (1), (B), (C), (2). It should also be noted that each of the listed modes has either a quasi-wavelength or a wave-like border.

Figure 131 shows the optical images of both longitudinal and cross sections of the contact boundary for modes (A), (B), (C). Figure 132 shows the same set of images, but for modes (1), (2), (3). In the case of an almost homogeneous pair, the difference in colour

Table 12. Amplitude and wave period values for cupronickel compounds

Mode	Wave amplitude A, μm	Wave period λ, μm
(A)	230 ± 5	530 ± 7
(B)	87 ± 7	212 ± 11
(C)	67 ± 6	190 ± 10
(1)	110 ± 5	268 ± 6
(2)	35 ± 9	100 ± 14

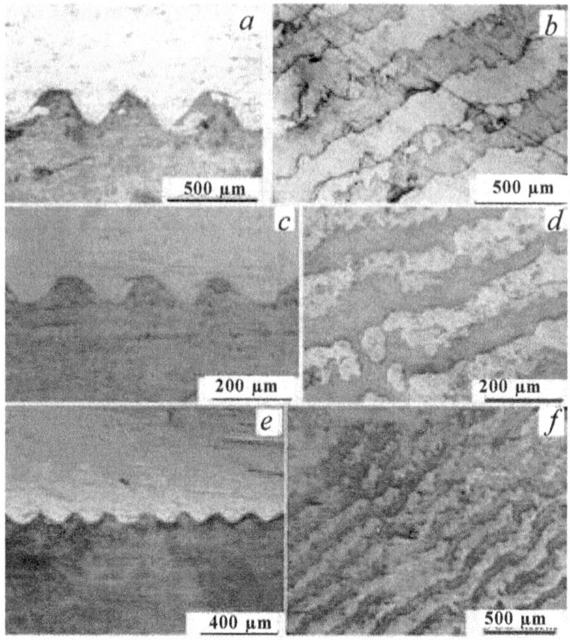

Fig. 131. Optical images of the interface. Cross section: *a* – (A), *c* – (B), *e* – (C); longitudinal section: *b* – (A), *d* – (B), *f* – (C).

between copper and melchior reflects the optical image. In the images below, the light colour corresponds to melchior, the dark colour corresponds to copper, and the third, intermediate colour, to melt.

For all the studied modes, cusps are observed at the interface (Figs. 131 and 132). They are clearly visible on the longitudinal sections, which are a set of alternating bands corresponding to different materials. There are also numerous local melting zones that differ in colour from the starting materials. The cusps, as well as local melting zones, form the fine structure of the contact boundary, which provides dissipation of the input energy. This behaviour was also characteristic of the Cu–Ta, Cu–Ti welded joints.

It should be emphasized that mode (3) has a flat interface (Fig. 132, *e*, *f*) and therefore it is impossible for it to determine the amplitude and period of the wave. In Fig. 132 *e* it is also seen that the surface of the sample (mode (3)) is strongly cut and covered with separate cusps. From these data, we can conclude that the surface is covered with splashes similar to those observed for Cu–Ta and Cu–Ti welded joints.

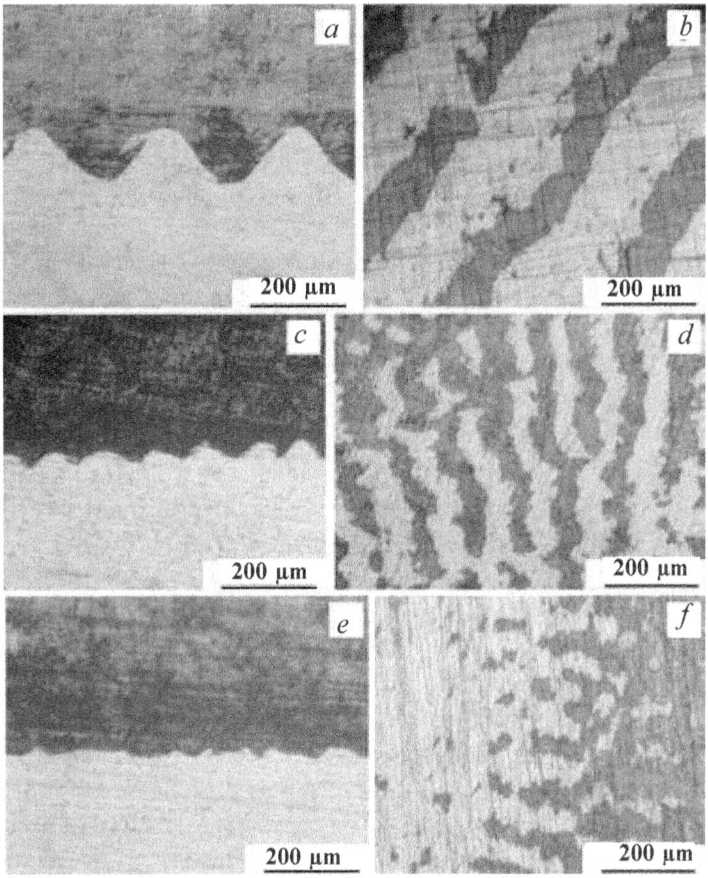

Fig. 132. Optical images of the interface. Cross section: *a* – (1), *c* – (2), *e* – (3); longitudinal section: *b* – (1), *d* – (2), *f* – (3).

The most energy was supplied to sample (2). In the longitudinal section (Fig. 132, d), bands of cupronickel are already visible. However, the mode is unstable, the bands often close, passing one into another. This behaviour of the transition zone has already been described for the Cu–Ti welded joint; it is radically different in the case of the Cu–Ta welded joint. At this stage, in the copper–melchior welded joint, the transition from splashes to waves occurs. The result of further intensification of the welding mode are the modes (B), (C), which are quite close to each other. For both cases, the transitional boundary is a wave, despite the fact that it is still cut, and there are separate cusps on its entire surface (Fig. 131, *d, f*). Next are the modes (A) and (1) (Fig. 131 *b* and Fig. 132 *b*, respectively), for which the interface is strictly wave.

For the welded joints (1), (2), (3) in Fig. 132 we note a clear pattern in the case of a longitudinal slice. It is seen that with an increase in the intensity of the welding mode, the irregularity of the interface decreases. For welded joint (3), in the case of individual islands, a pronounced surface roughness is observed, while for joint (1) rather smooth bands are observed. Such behaviour fits well with the findings obtained earlier, where the rule was formulated: "when the welding regime is intensified, the roughness (fractal dimension) drops". Quantitative calculation of this dependence will be made in the next section.

Modes (D) and (3) are located near the LB window of weldability. For mode (D), welding took place on a certain area and it seemed that the connection was obtained. However, when grinding a sample, the welded joint exfoliated. And vice versa: the welded joint obtained by mode (3) was not stratified. Thus, it can be argued that the LB passes somewhat higher in the case of mode (D) and lower for mode (3) (Fig. 130).

The final intensity distribution for all the studied modes is as follows: (A), (1), (B), (C), (2), (3), (D). The flat boundary is characteristic of the welded joints (D), (3), the quasi-wave boundary is (2), (C), (B); wave-like border – (1), (A).

At the end of this section, we would like to emphasize several interesting features observed for copper–melchior welded joints. As already noted, the mode (D) is below the LB of the weldability window. However, it turned out that in some areas the welding of copper and melchior did occur (Fig. 133). Visible dark spots (marked in red), corresponding to the areas where welding occurred.

For copper–melchior welded joint, vortex formation was observed in the solid phase. In Fig. 134 there are visible unfinished large-scale turns (macro turns) in melchior (shown by white arrows). Here, SEM images were used, since such thin structural elements, like vortex formation, cannot be detected by optical methods.

As mentioned above, numerous copper melting zones were discovered for the copper–nickel welded joints, which are a dilute nickel solid solution in copper. Such a solution is formed from melchior containing 19 at.% Ni. Since one of the joined plates is pure copper, the solution contains only 7 at.% Ni. In the case of a copper–melchior welded joint, no vortices were found inside the local melting zone, in contrast to the Cu–Ti welded joints.

Fig. 133. SEM image of the surface of the cupronickel after explosive welding, mode (D).

9.5. Fractal description of the interface for the copper-melchior alloy welded joint

For a homogeneous melchior welded joint, the fractal analysis, namely the 'coastline' method, was used to determine the surface roughness of the interface.

The analyzed surfaces were first selected from the first group (samples (A), (B), (C)), and then from the second group (samples (1), (2), (3)). It should be noted that this choice is justified by a strict decrease in the input energy. Equal amplitudes for different samples are also absent (Table 12). Therefore, the pattern: "with the intensification of the welding mode, the fractal dimension decreases" [106], should not have any exceptions.

Figures 135, *a, c, e* are the longitudinal section for the welds (A), (B), (C), respectively, and in Fig. 135, *b, d, f* – the same longitudinal section, prepared for the calculation of the fractal dimension. In Fig. 136, *a, c, d* are represented by a longitudinal section for the welded joints (1), (2), (3) respectively, and in Fig. 136, *b, e, f* – the same longitudinal section, but two-colour, designed to calculate the fractal dimension.

After calculations, the following fractal dimension for the set of the (A), (B), (C) welded joints was obtained: $D(A) = 1.01 \pm 0.01$,

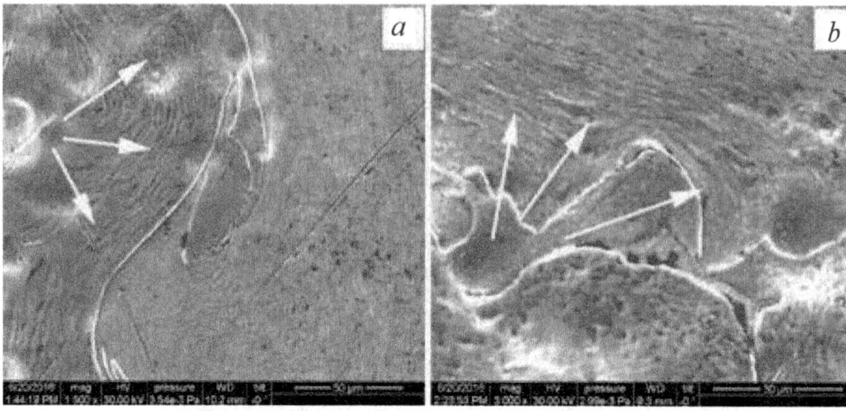

Fig. 134. SEM image of macroturns (indicated by arrows) in cupronickel alloy (*a*) and copper (*b*); mode (B).

D(B) = 1.05 ± 0.02, *D*(C) = 1.16 ± 0.02. Immediately striking is the fact that the value of the dimension for the (A) welded joint is close to unity. It can be concluded that this surface has an extremely weak irregularity, which does not change the fractal dimension. The (B) welded joint also has an extremely low dimension. The obtained data well reflect the absence of roughness of the relief of the welded joints of the samples (A), (B) (Fig. 135 *a, c*). With a further decrease in the input energy, the fractal dimension grows (sample (C)), which corresponds to similar situations for the Cu–Ta welded joint.

For the set of the (1), (2), (3) welded joints, the following values of the fractal dimension were obtained: *D*(1) = 1.01 ± 0.02, *D*(2) = 1.20± 0.03, *D*(3) = 1.24 ± 0.02. Welded joint (1), as well as welded joint (A), has a fractal dimension close to unity. Figure 136 *a* presents the surface of the longitudinal slice for this welded joint, whence it is seen that the relief is very smooth. A further decrease in the energy input leads to an increase in the fractal dimension, as in the previous cases.

The final intensity distribution for all the investigated copper-melchior welded joint modes obtained by explosive welding, as already mentioned above, is as follows: (A), (1), (B), (C), (2), (3), (D). The obtained regularity strictly coincides with the dependence for the fractal dimension (from the point of view of the distribution of welding modes over the input energy) [106].

At the end of this section, it should be noted that the fractal dimensions obtained for the copper–titanium welded joints [106] are, on average, higher than for the copper–melchior welded joints.

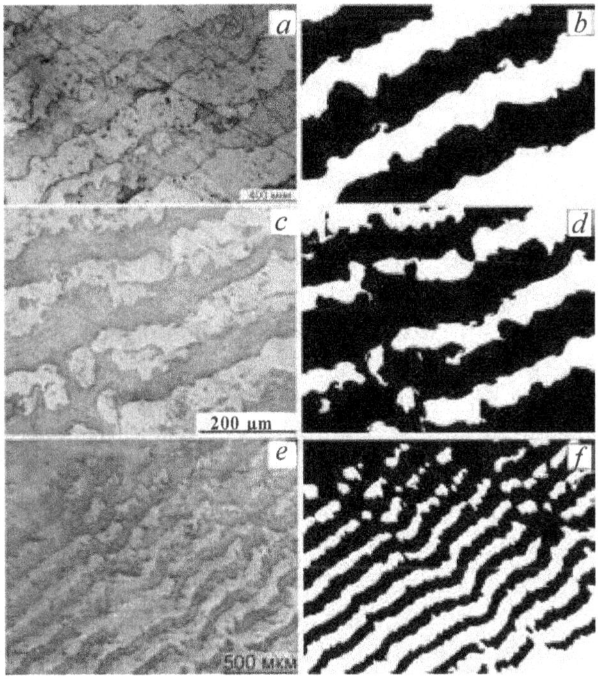

Fig. 135. Longitudinal section: *a, b* – welded joint (A), *c, d* – (B), *e, f* – (C).

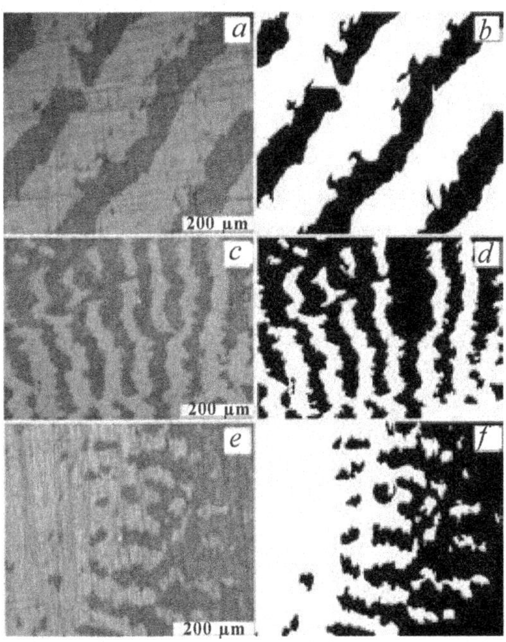

Fig. 136. Longitudinal sections: *a,b* – welded joint (1); *c, d* (2); *e, f* (3).

The reason for this pattern may be due to the fact that the resulting interfaces for homogeneous pairs are smoother. In a homogeneous pair, the hardness of the flyer metal coincides with the hardness of the substrate; therefore, less energy is 'spent' on such an element of the fine structure as cusps.

Conclusions to chapter 9

1. The evolution of the interface surface of a conventionally homogeneous copper–melchior welded joint has been investigated during the intensification of the welding mode. The similar behaviour of the surface relief with such welded pairs as copper–tantalum, copper–titanium was revealed. Typical examples of a fine structure for a copper–melchior welded joint were found: cusps, groups of cusps uniting into splashes, local melting zones, and vortex formation on the surface of copper and melchior alloy.

2. As in the case of dissimilar pairs, a quasi-wave interface surface was found for the copper–melchior welded joint. In the case of a longitudinal section, this surface contains imperfect copper strips, which pass into each other. Such a structure is close to the quasi-wave interface, observed in the case of a copper–titanium welded joint, and qualitatively differs from the quasi-wave boundary for a copper–tantalum welded joint.

3. With an increase in the intensity of explosive welding for copper–melchior welded joint, a decrease in the irregularity of the transition zone is observed. The following regularity is established for the copper–melchior welded joint: the maximum fractal dimension is observed for a flat boundary, and with a further intensification of the welding regime, the fractal dimension begins to fall. For the copper–melchior welded joint, the wave-like boundary, starting from certain values of the energy supplied, ceases to be a fractal object.

4. Local melting zones for a homogeneous pair are a dilute solid solution of nickel in copper. No vortices were found inside the local melting zone, in contrast to the copper–titanium welded joints.

10

Structure of multilayer composites produced by explosive welding

The results of the study obtained by explosive welding of multilayer composites based on steel and magnesium, as well as welded joints Nb–Cu, Ta–Cu. The role of intermediate layers playing the role of buffers is revealed. In some cases, the introduction of an intermediate layer between the functional layers, preventing the formation of intermetallic compounds, promotes the weldability of these materials. Ti–Al–Mg composites were studied using a computer simulation process. The distribution of materials and contours of speed, pressure and effective plastic deformation are shown for some moment in the process of simulation. The cumulative jet is also reproduced during the simulation. For magnesium based composites, the possibility of magnesium boiling and foaming is shown. Consequences to which the presence of foam can lead to during cooling are revealed. The presence of ultrafine-grained refractory (Nb or Ta) and Cu phases was detected for the Nb–Ta and Ta–Cu welds. For the Ta–Cu welds, the presence of a non-equilibrium phase based on Ta or quasicrystals in the Cu matrix is also detected. The coexistence of two ultrafine-grained phases causes a higher hardness of the intermediate layers and a higher yield stress of the joints. A comparison is made with the Cu–Ta jointinvestigated above.

Particular attention is paid to the microstructure of the Cu-Ta multilayer composite produced by explosive welding. Copper-tantalum compounds with different interfaces (flat, quasi-wave, wave) are studied. The role of mutual solubility of the initial elements on the formation of the structure of the welded joint has been revealed. Comparison of multilayer composites with bilayer composites, as well as with compounds obtained using torsion under pressure, has been carried out. A significant difference in microstructures and the causes of differences were revealed. Electron-microscopic studies of the evolution of the microstructure of metals, their similarity and difference for composites of several types were carried out.

10.1. Structure and properties of certain composites

This section presents the results of the study of composites produced by explosive welding; based on steel [107, 108], three-layer composite Ti–Al–Mg [109, 110], two-layer composites, Nb–Cu, Ta–Cu [113].

10.1.1. Steel-based composites

In work [107], high-strength maraging steel ZI90-VI, dural, beryllium bronze, and titanium alloy OT4-1 were used. Welded packages of dissimilar materials: three and five alternating plates. The following welding parameters were used: impact angle of 20°, detonation speed of 2450 m/s, impact velocity of 850 m/s. Materials before welding were subjected to appropriate annealing.

Three-layer composite of dural and steel. The interface is flat, the surface roughness is practically not observed. The selected modes correspond to the lower boundary of the weldability window. The microhardness of duralumin after welding is approximately 200 HV10, which is higher than in the initial state (77 HV10). The microhardness of steel ZI90-VI also increased and amounted to 500 HV10, which exceeds the initial value of 350 HV10. Composite samples were destroyed after 5–7 alternating bends without stratification, the fracture mechanism is viscous.

Five-layer composite of dural and steel. Section surfaces are almost flat, minor irregularities. On both sides of the interface, there is a transition zone containing small islands of supposedly intermetallic inclusions. The microhardness of materials in a five-layer composite is the same as in a three-layer composite. Composite samples were destroyed after 7–10 alternating bends without stratification, the fracture mechanism was viscous, patched.

Five-layer composite of beryllium bronze and steel. Figure 137 is an SEM image of the cross section of this five-layer composite.

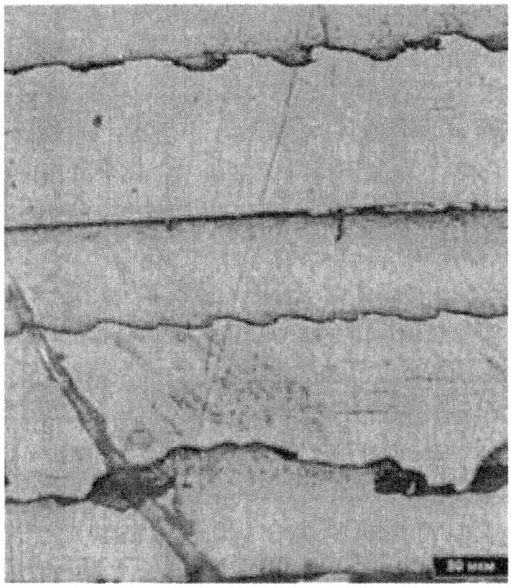

Fig. 137. Cross-section of the interfaces between the five-layer composite made of BrB2 alloy and ZI90-VI steel [107].

Wave-like surfaces are observed, near them a zone of partial reflow. Numerous cusps on the surface of the section belong to a harder material – steel. In [107], it is noted that due to the short duration of the explosive welding process, it is not possible to obtain intense mixing in the transition zones due to diffusion movements of atoms over such long distances. A similar point of view was expressed in previous chapters.

The microhardness of bronze in the composite is 300 HV10, which is slightly higher than the original. Microhardness of steel, as in the previously reviewed composites, reaches 500 HV10. In the bend test no stratification was observed. The number of alternating bends before failure was 10–12.

Five-layer composite of steel, alloy OT4-1 and dural. There is a weak wave-like shape of the boundary. In the contact zone, areas of altered chemical composition are observed. The microhardness of duralumin is 200 HV10, titanium alloy 300 HV10, steel ZI90-VI 500 HV10. The number of alternating bends was 7–10.

A study of steel-based composites has shown that cusps are observed on wavy interfaces, but not on flat surfaces. These results contradict the results previously obtained for the Cu–Ta welded joint. Layered metal-based composites with a sandwich structure have a unique combination of physical and mechanical properties, as well as increased resistance to brittle fracture over a wide temperature range. As a result, layered metal composites of structural and functional purpose are becoming more and more widely used in various branches of transport and power engineering, shipbuilding and aerospace engineering.

The constructions of modern layered metal composite materials (LMCM) provide the technological possibility of combining high-strength and ductile components, which allows controlling the mechanical properties of a multilayer material by changing the volume fraction and the order of alternation of layers. In [108], a set of mechanical properties and characteristics of crack resistance under various loading conditions of LMCM based on 09G2S and EP678 steels with different structure dispersion in the state immediately after explosive welding, as well as after additional heat treatment, was determined. Three initial plates of structural carbon steel 09G2S and two plates of maraging steel EP678 were used as initial components of the 5-layer composites. Both wavy and rectilinear boundaries form in these joints.

Heat treatment. The explosive layered composites were subjected to additional heat treatment at 500°C for 3 hours. The purpose of heat treatment of the finished laminated composites is to achieve the maximum level of the strength properties while maintaining the high toughness of the composite, which can be achieved by hardening the steel EP678. The heat treatment carried out after explosive welding does not cause any significant activation of diffusion processes at the interlayer boundaries. Based on the data obtained, it was concluded that the heat treatment of the LMCM of this composition allows one to adjust the ratio of the strength properties, providing both strengthening and softening effect on the individual components of the laminated material.

Tensile tests. The highest level of strength properties is achieved in welded composites that have an UFG (ultrafine-grained) material material in the interlayer, without conducting subsequent heat treatment of the composite. Additional heat treatment does not have a significant hardening in the layers of UFG EP678 steel. The hardening effect due to the ageing of layers of MSS (maraging steels) is most noticeable on layered composites with EP678 FG (fine-grained) steel, while the plastic properties are somewhat reduced.

Impact tests. An anomalous effect of increasing the characteristics of the impact toughness of composite 1 (UFG) immediately after explosive welding has been revealed. For composite 2 (FG), lowering the test temperature from +20 to –60°C leads to complete destruction of the sample. During impact tests, a slight deviation of the crack trajectory along the interface with the formation of splits of cracks of various lengths occurs, which contributes to the inhibition of the crack and increase the characteristics of impact strength.

10.1.2. Magnesium-based composites

Let us turn to the study of magnesium-based composites. Magnesium alloys have a great advantage as light alloys, and in addition, due to their wide use [59].

However, their use is limited due to the low corrosion resistance of Mg. In this regard, the high corrosion resistance of Ti could expand the range of application of Mg alloys, but the Ti–Mg binary phase does not exist because of the difference in the crystal lattices. At the same time, Ti–Al and Mg–Al can be successfully obtained. Therefore, the idea of introducing an Al interlayer between Ti and Mg plates is fruitful for creating new structural materials and multilayer Ti–Al–Mg laminates [109].

Experimental model, plates used: Ti alloy, Al alloy, AZ31B Mg alloy. A wave boundary between Al and Mg layers and a flat boundary between Al and Ti layers were found. In addition, a computer simulation process was implemented. The model for three plates is shown schematically during explosive welding in Fig. 138, *a* and at simulation on Fig. 138 *b*. The distribution of materials and the contours of velocity, pressure, and effective plastic deformation are shown for some point in the simulation process (Fig. 138 *b*). Analysis of the simulation process demonstrates wave morphology at the Al–Mg interface and flat morphology at the Al–Ti interface. The cumulative jet is also reproduced during the simulation. Most jets emanate from the base (Mg) plate. The reason is that the density of the Mg alloys is lower than that of the Ti and Al alloys.

Discussion. A prerequisite for explosive welding is the formation of a cumulative jet, which is essential for good weldability. From the point of view of the analysis of forces, it is necessary that the surfaces come into contact sufficiently close, in the range of action of the forces of attraction. The forces of attraction and repulsion are equal at some equilibrium distance. Potential energy must reach a minimum value in order to overcome repulsive forces at the atomic level. The distance between the surfaces should be small enough to ensure the action of gravity. And then the clutch occurs. The cumulative jet has a remarkable effect on the surface of the cleaned

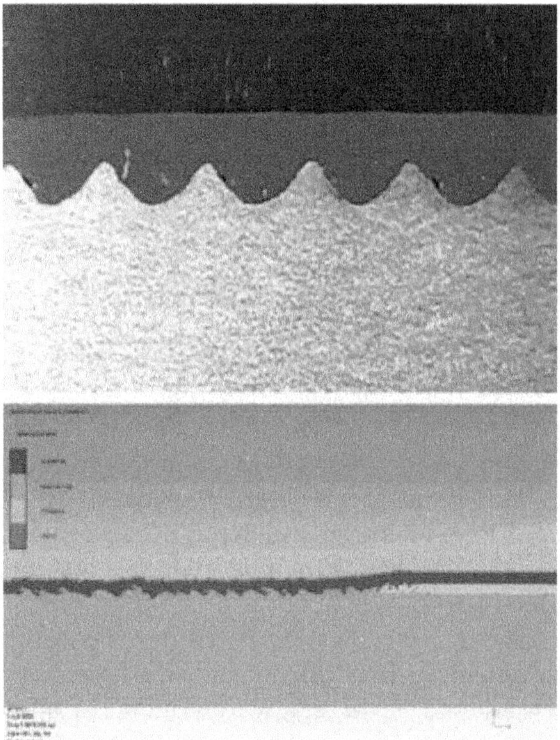

Fig. 138. Cross section of a three-layer composite: in experiment (*a*); during simulation (*b*)

plates, which is successful in obtaining the necessary adhesion in the experiment. When observing the entire simulation process, most of the jets emanate from the base (Mg) plate, whereas they do not occur between layers 1 and 2. This agrees with the theoretical results that the wave surface is always along the jet, but the jet does not occur in the case of an almost flat surface.

Conclusion. During the simulation, it was possible to obtain a wave morphology on the Al–Mg surface and an almost flat morphology on the Ti–Al surface, which agrees with the morphology observed in the experiment.

Similar work was done in [110]. The structure of interfaces in Mg–Ti composites produced by explosive welding was investigated. The material for the study was the multilayer composite materials AMg6–Al–Ti–Ma2–1 and Ti–Al–Ti–Mg. Here, AMg6 is an aluminium alloy, and Ma2–1 is a magnesium alloy containing 92.9% Mg, 3.8% Al, 0.6% Zn, 0.2% Mn.

Buffer layers. When creating composite materials, intermediate layers are used, which either play the role of a buffer, separating metals due to the small stock of their plasticity, or a diffusion barrier, preventing the formation of intermetallic compounds.

When creating composite materials, aluminium–magnesium oxide films on the surface of both magnesium and aluminium interfere with the weldability of these materials. One possible solution to this problem is the introduction of a layer between Al and Mg. But with the direct connection of these materials in the heat-affected zone, a large number of intermetallic compounds occur. Then it is proposed to introduce a titanium intermediate layer between the functional layers of aluminium and magnesium.

The structure of the boundary between the magnesium alloy and titanium in multilayer composite materials was investigated in [110]. An attempt was made to answer the question: is it advisable to introduce a titanium intermediate layer as a buffer layer between aluminium and magnesium?

The magnesium–titanium metal system is a system with no mutual solubility. Nevertheless, it has been established experimentally that the solubility of titanium in liquid magnesium at temperatures from 650°C to 1500°C is 0.012 and 1.03%, respectively [111].

The solubility of titanium in solid magnesium varies from 0.08% at 350°C to 0.19% at 650°C. In the magnesium–titanium system, intermediate intermetallic phases are not formed. A solid solution of magnesium in titanium containing up to 1.32% Mg is in equilibrium with liquid magnesium.

In the MA2-1–Ti–Al–AMg6 and MA2–1–Ti–Al–Ti composites, the intermediate layer (Ti–Al) performs two functions. On the one hand, titanium, having limited mutual solubility with magnesium, is a diffusion barrier between magnesium and an aluminium alloy. On the other hand, aluminium in this composition acts as a plasticity buffer between hard-to-weld titanium and AMg6 aluminium alloy. The resulting four-layer composite has high strength properties, even after prolonged temperature testing.

Magnesium boils and foam formation. The welding of magnesium and its alloys is characterized by the formation of cracks in the weld. Together with extensive pore formation, hot cracks are the main problem for any type of magnesium welding. Experimental observation of pores in the welded joint zone may be due to the evaporation of magnesium [112].

Figure 139 shows the structure of the Ti–Mg interface in the AMg6–Al–Ti–Ma2-1 composite material (SEM, cross section). The internal structure of the local melting zone includes titanium particles along with fairly large fragments of titanium of arbitrary shape with a size of 5–15 μm.

Since titanium and magnesium are not soluble in each other, at certain temperatures and pressures during the explosive welding process, the formation of a titanium–magnesium colloidal system is possible. We give the values of the boiling points and melting points of both metals: T_{boil} (Ti) = 3560 K, T_{melt} (Ti) = 1933 K, T_{boil} (Mg) = 1363 K, T_{melt} (Mg) = 922 K. Due to low boiling point of Mg we can expect the formation of gas and liquid magnesium phases near the interface in the process of explosive welding.

It is considered established that the formation of any stable foam in a pure liquid is impossible. In those zones where both metals are melted, foam formation is possible, in which liquid interlayers are formed of a magnesium–titanium colloidal solution, and the bubbles

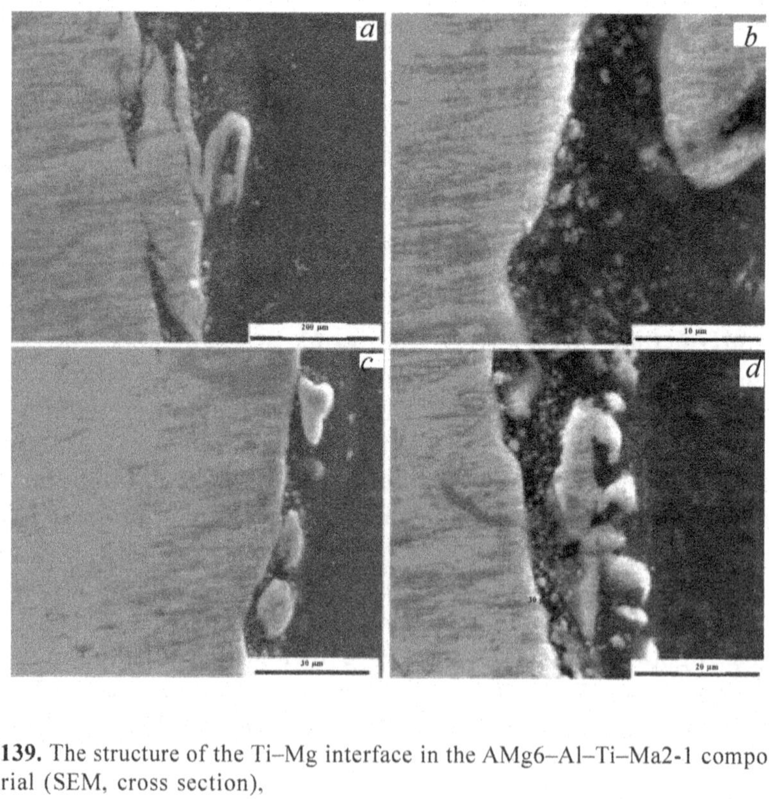

Fig. 139. The structure of the Ti–Mg interface in the AMg6–Al–Ti–Ma2-1 composite material (SEM, cross section),

are filled with magnesium vapor [112]. Voids and pores are clearly visible on the electron microscopic image of a longitudinal section of the interphase boundary between titanium and the magnesium alloy, .

Beating is necessary for the formation of foam, shaking the liquid, which is provided during the explosion. Due to the ultrafast quenching, the foam can persist when cured. Then pores will be observed near the interface.

Zones of risk for the studied composites are zones of molten magnesium in the event that they turn out to be, moreover, zones of boiling of magnesium.

Findings. As a result of the introduction of a buffer titanium layer in the MA2–1–Ti–Al–Ti composite, the boundary between titanium and magnesium is loose, porosity is observed, and the continuity of the detachable weld is broken. Thus, the boundary under study is a weak link in the whole structure. In the MA2–1–Ti–Al–AMg6 composite, the electron microscopic defects of the internal structure of the interphase boundary between titanium and a magnesium alloy are not detected.

10.1.3. Nb–Cu and Ta–Cu welded joints

A related paper [113] is devoted to an analysis of the advantages that the intermediate layer (IL) provides for Nb–Cu and Ta–Cu welds obtained by explosive welding. Figure 140 shows the SEM images of the interfaces for Nb–Cu (*a*), (*c*), (*d*) and Ta–Cu (*b*), (*e*), (*f*) composites made by welding using an explosion at a high HCV collision velocity HCV (*a*), (*b*) and low collision velocity LCV (*c*), (*e*).

Main results. The intermediate layer increases the bending strength of Nb–Cu and Ta–Cu welded joints, without losing a large elongation.

The coexistence of ultrafine-grained refractory metals and copper phases increases the hardness of the intermediate layer, which leads to a high bending stress.

For the first time, Ta–Cu-based intermetallic compounds and dodecahedral Ta–Cu-based quasicrystalline phases were detected in a copper matrix.

The formation of an intermediate layer results in a less deformed plate of the refractory metal near the interface.

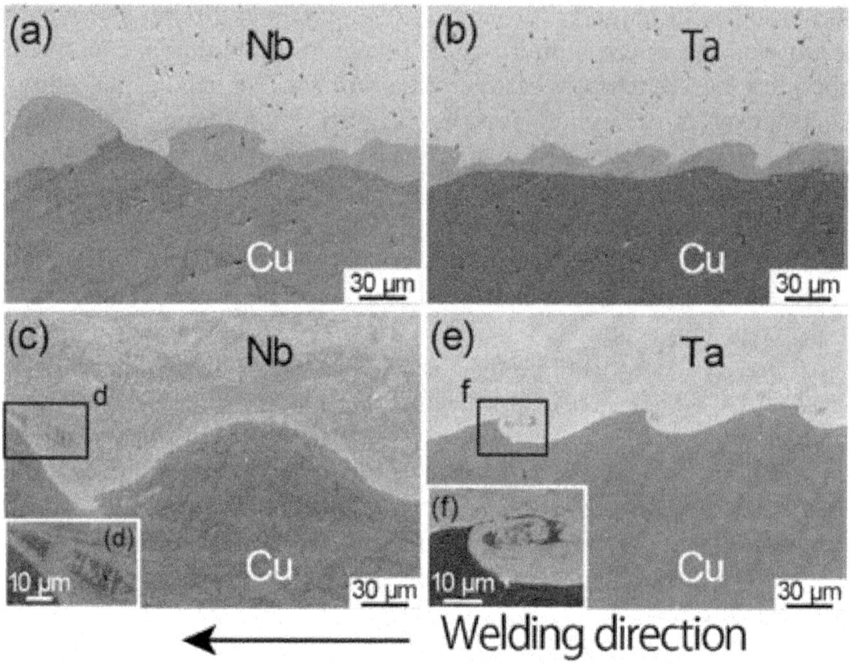

Fig. 140. SEM images of interfaces for composites Nb – Cu (a), (c), (d) and Ta – Cu (b), (e), (f) obtained by explosive welding at HCV (a), (b) and LCV (c), (e).

The hazard of the shock wave for the base metal plates was reduced, due to the formation of an intermediate layer at the interface for Nb–Cu compounds.

Deformation behaviour without an intermediate IL layer and in the presence of IL. A systematic comparative analysis was performed to quantify the flexural strength of wave surfaces with IL and without IL in Nb–Cu and Ta–Cu welded joints, which were obtained at high (HCV) and low (LCV) collision velocities during explosive welding. The deformation curves for all welded joints were similar, but in the presence of IL, a higher bending stress was observed than in the absence of IL. Moreover, cracks were observed in the Nb plate for Nb–Cu welds without IL, but were not observed in the presence of IL.

TEM observations confirmed the existence of ultrafine-grained refractory metals and copper phases in ILs interlayers. In addition, Ta–Cu based intermetallic welded joints and dodecahedral quasicrystals, which are a non-equilibrium phase, were also observed simultaneously with the existence of the Ta and Cu phases in ILs.

HCV weldability is supported by less deformation of the plates of refractory metals near the interface than in the case of LCV, due to the formation of ILs. This helps to achieve higher hardness and, accordingly, to obtain compounds with high flexural strength while maintaining a large elongation

Conclusion. A comparative study of wave surfaces with or without intermediate layers for Nb–Cu and Ta–Cu welded joints obtained for the HCV and LCV modes was used to understand which of the surfaces provides better flexural strength and deformation behaviour in bending experiments.

Important facts were discovered:

1) The Nb–Cu and Ta–Cu welded joints, having wave surfaces with an intermediate scrapping of IL formed in HCV, had a higher yield strength in bending than that observed without IL at LCV. A higher yield stress was associated with ILs, which have a much higher hardness compared to base metals.

2) Interlayers of ILs compounds detect the presence of ultrafine-grained refractory (Nb or Ta) and Cu phases. In the IL for Ta–Cu welded joints, the presence of a non-equilibrium phase based on Ta or quasicrystals in the Cu matrix is also detected. The coexistence of two ultrafine phases is responsible for the higher IL hardness and higher yield stress of the welded joints.

3) No cracks were observed on Nb plates in the Nb–Cu welded joints having wave surfaces with IL, but cracks were observed without IL. Weldability with HCV is associated with less deformation of refractory plates near the interface than at LCV, thanks to the work of the formation of IL.

The papers [10, 58] and also the authors of the monograph [114] are cited in [113].

Comparison of results [113] with our results [10, 58, 114]. In article [113] it is considered that both Nb and Ta melt during explosive welding.

But we in the cited articles believed that tantalum does not melt and local melting zones are zones formed by the copper melt (then frozen), and tantalum particles do not melt. We believe that the wavy interface is filled with local melting zones, cusps (splashes) and areas with a quasi-wave structure. These concepts differ significantly from those developed in [113], where the wave surface is also considered, but there are no cusps (splashes) or a quasi-wave structure.

On the other hand, we did not observe tantalum-based quasicrystals. The article [113] does not define quasicrystals (dodecahedral symmetry). Moreover, the works of Nobel Laureate Dan Shechtman, who discovered quasi-monocrystals of fivefold symmetry, are not even mentioned. There is also no definition of an intermediate layer (IL). Perhaps the IL corresponds to the transition zone in our articles, but there is no such comparison. The article [113] assumes that solidification of the melt (solidification from the melt) can be explained with the help of the granulating fragmentation (GFR) proposed by us. GFR is discussed in detail in Chapter 4 (Section 4.1).

But the formation and dispersion of quasicrystals (dodecahedral symmetry) from the tantalum layer with a BCC lattice is also in doubt. It also remains unclear how intermetallic compounds appear in the Ta–Cu system, given the absence of mutual solubility.

10.2. Multi-layered composites based on Cu–Ta

The microstructure of the Cu–Ta welded joint (two-layer composite), as already discussed in previous chapters, when the welding mode is intensified follows the following rules: the transition from the region below the lower boundary (LB) of the weldability window directly to LB is accompanied by the consolidation of individual cusps into regularly distributed splashes with random directions. A further increase in the amount of energy supplied to the system leads to the formation of a quasi-wave surface, on which splashes simultaneously coexist, as well as multidirectional waves with different amplitudes. The final stage is the formation of a wave-like boundary in the centre of the weldability window. The main elements of the topography of the interface, thus, are the cusps, splashes, wave- and quasi-wave surfaces. In this section, an attempt is made to find out what relief changes occur during the formation of a Cu–Ta-based multilayer composite during explosive welding. In addition, a comparison is made with the microstructure of a similar composite, obtained by torsion under pressure.

10.2.1. Experimental material and procedure

Explosive welding was carried out by the Volgograd State Technical University, OJSC Ural Chemical Engineering Plant (Yekaterinburg). Parallel arrangement of the plates was used. An explosive charge was placed on the upper (cladding) plate.

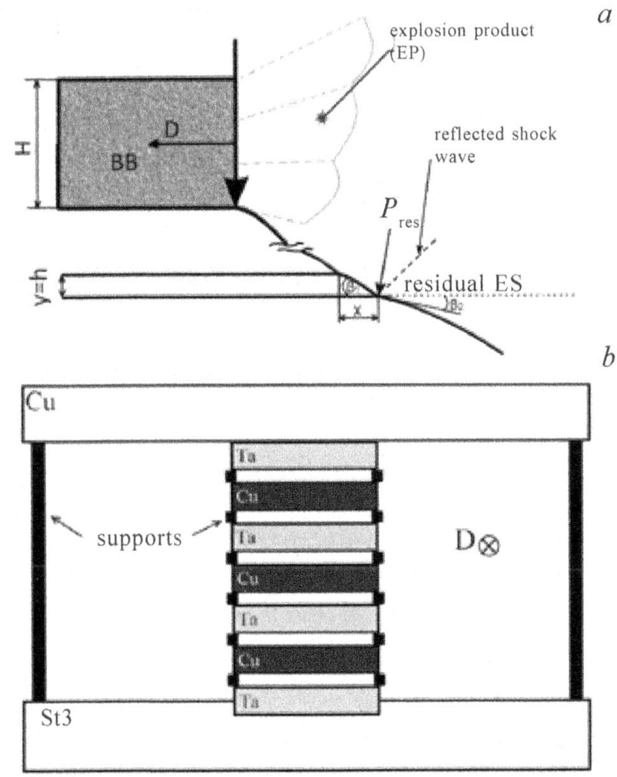

Fig. 141. Diagram of transverse collisions of plates at their multilayer throwing: *a* – parallel to the movement of the detonation wave, *b* – perpendicular to the movement of the detonation wave.

The main part of the work was the study of a multilayer composite, where copper and tantalum, which do not have mutual solubility, were chosen as the main starting materials. A comparative analysis was also carried out with a copper–tantalum-based two-layer composite having the following welding mode parameters: $\mathbf{C}_p\downarrow$ ($\gamma = 3.8°$, $V_{con} = 2564$ m/s), $\mathbf{C}_p$ ($\gamma = 5.2°$, $V_{con} = 2571$ m/s) and $\mathbf{C}_w$ ($\gamma = 11.8°$, $V_{con} = 2140$ m/s), where γ is the impact angle, V_{con} is the velocity of the contact point. The subscript corresponds to the shape of the boundary – flat (plain), wave (wave). Welding with parameters below the lower boundary of the weldability window is designated as ($\mathbf{C}_p\downarrow$).

The multilayer composite was obtained using a scheme of several successively colliding plates (Fig. 141 *a*). This figure shows a schematic cross section of a multilayer composite parallel to the movement of the detonation wave (DW) during the initiation of the explosive [115]. We direct the x axis along the initial position of the striker in the direction opposite to the motion of the detonation front. The y axis is in the direction of the plates offset. In Fig. 141 *a* the angle β_1 is the angle of rotation by the distance y, and the angle β_2 is the angle of rotation after the collision calculated from the law of conservation of momentum. P_{exp} is the pressure of the explosion products on the plate at a distance y, taking into account the impact.

Figure 141 *b* is a schematic cross section of a multilayer composite perpendicular to the movement of the detonation wave (D). This figure shows exactly how the sample was obtained. At the initiation of explosives under the action of the DW, the flyer copper plate breaks the struts and, together with the tantalum plate, falls onto the next copper plate, forming the first interface (1). Similarly, six studied interfaces were obtained. They correspond to the following parameters: the velocity of the contact point for all is the same V_{con} = 2400 m/s;

(1) – γ = 10.63°, E_C/E_e = 0.01098;
(2) – γ = 11.23°, E_C/E_e = 0.01523;
(3) – γ = 11.01°, E_C/E_e = 0.01250;
(4) – γ = 10.96°, E_C/E_e = 0.01533;
(5) – γ = 10.86°, E_C/E_e = 0.012256;
(6) – γ = 10.77°, E_C/E_e = 0.07287.

E_C is the collision energy (the difference in kinetic energy before and after the collision), and E_e is the total energy of the explosive. It is important to note that E_C/E_e has similar values for the interfaces (1) – (5). However, for boundary (6), this value is seven times larger, which is not surprising, since the last plate of tantalum is fixed and it accounts for the energy of the entire upper packet.

Metallographic analysis was performed on an Epiquant optical microscope equipped with a SIAMS computer system. The study of the microstructure was performed using JEM 200CX transmission electron microscopes and a Quanta 200 3D scanning electron microscope.

10.2.2. Microstructure of Cu–Ta multilayer composite materials produced by explosive welding

Obtaining and studying the structure of multilayer composites on the basis of Cu–Ta is actually the purpose of this study. Figure 142, show the cross-section of the Cu–Ta multilayer composite. The following

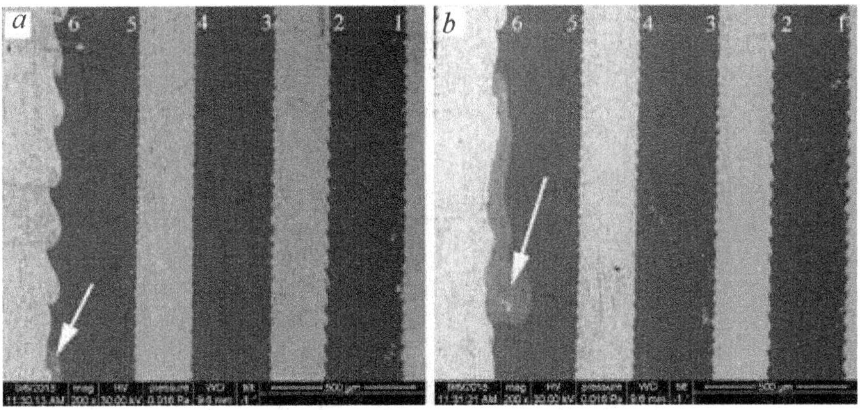

Fig. 142. The cross section of multilayer composites copper tantalum: *a*, *b* – different places of the sample.

transition zones were studied: (1), (2), (3), (4), (5), (6). As can be seen from Fig. 142 *a*, they all have a wave-like boundary. Zones (1)–(5) have approximately the same input energy, and therefore, a similar interface. The transition zone (6) is formed by the coupling of the penultimate and last (fixed) plate, so it accounts for the energy of the entire upper package. Zone (6) has a maximum energy input, approximately seven times greater than all other zones. Therefore, on its surface one could expect a maximum amplitude and period of the wave, as observed in Fig. 142 *a*. However, it should be noted that zone (6) is very unstable: in Fig. 142 *b* it can be seen that when moving along the composite, the structure of zone (6) ceases to be a pronounced wave, instead, a rather extensive melt section is observed (indicated by an arrow).

All of the above has already been observed by us for isolated welded joints. The logical step would be to compare the results obtained for a multilayer composite with those obtained earlier for a two-layer composite.

Similarity:

1) In the case of a multilayer composite, we did not obtain a flat interface, but in Fig. 143, where the relief of zone (6) is represented (copper is etched), characteristic relief elements are observed – splashes and waves. Thus, it can be argued that the transition zone (6) is a quasi-wave boundary. Both splashes and the quasi-wave boundary were considered by us earlier in the case of isolated copper–tantalum-based welded joints.

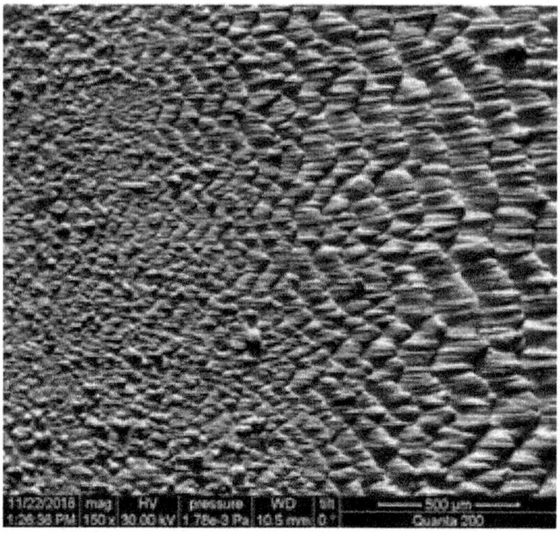

Fig. 143. The surface relief of tantalum in the case of zone (6), copper etched.

2) A smooth transition from splashes to waves, which is visible in Fig. 143 when moving from the left part of the figure to the right, it was observed by us for isolated welded joints.

3) In the structure of the transition zones, areas of local melting of the material were detected (Fig. 142, b), which is a characteristic feature of isolated copper–tantal-based welded joints.

Difference:

1) All six transition zones ((1)–(6)), as it turned out, have a quasi-wave boundary (Fig. 144, *a–f*). By itself, this fact is rather unusual. However, if we consider these compounds in more detail, we can find that zones (1)–(4) and to a lesser extent (5) have an already well-known quasi-wave boundary. One can see areas where a good wavy structure appears, but with different amplitude and wave period (for example, areas '2', '3' in Fig. 144 *a*), as well as parts of a flat boundary covered with separate splashes (for example, section '1' in Fig. 144 *a*). However, Fig. 144 *f* shows a quite smooth surface on which there is no roughness. For the transition zone (6), three states are realized simultaneously: a smooth surface, a flat boundary (consists of separate cusps), and a wavelike boundary. A similar relief for isolated welded joints was not observed.

2) Previously, we assumed that with the intensification of the welding regime, there is a transition from flat to quasi-wave, and then to a wavy boundary. The transition zone (6) has a higher energy input than (1)–(5). However, all zones correspond to a wavy boundary, and

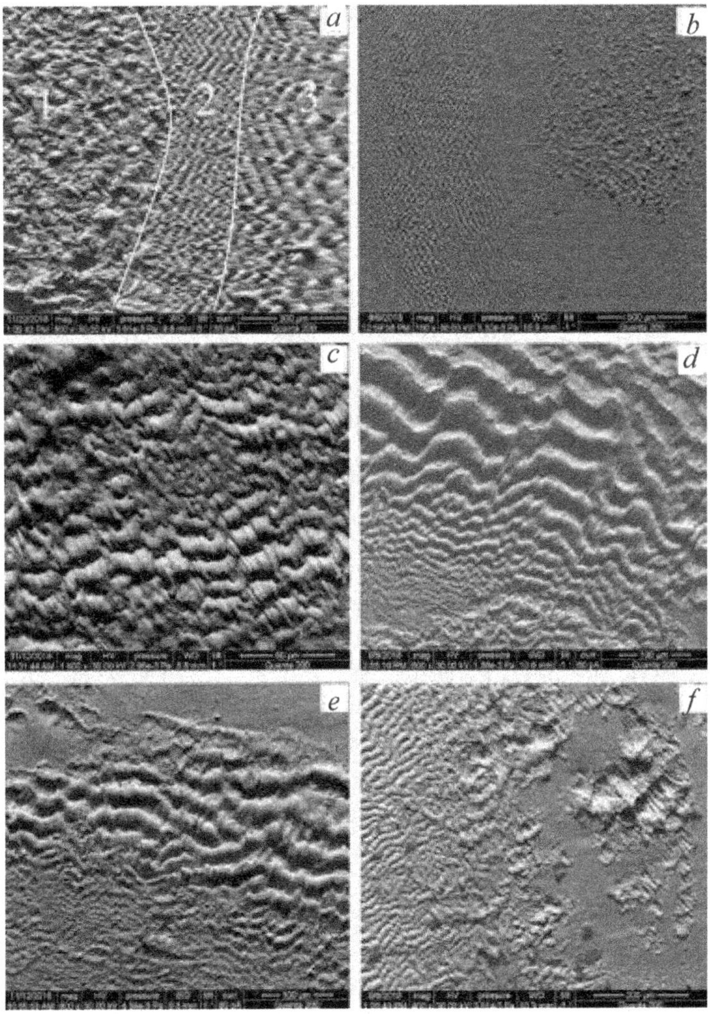

Fig. 144. Relief of the surface of tantalum (copper etched out): *a–f* – zones (1)–(6).

for (6) completely smooth sections were found. This pattern is also not typical for isolated welds. Such behaviour can be explained by the non-uniform energy distribution over the material surface in the case of the implementation of the multilayer flyer scheme of plates.

3) In the case of a multilayer composite, zones of local melting were found. However, the structure in the form of solidified particles of tantalum in molten copper was not observed.

10.2.3. Mechanical alloying in the case of torsion under pressure for the Cu–Ta system

As mentioned above, the composites were produced using explosive welding, and torsion under pressure. These methods have a number of common features: the studied processes run in open systems; the structure of the compound in both cases is formed with a strong external influence; non-linear effects are important; the final structure is subject to the principles of self-organization. It should also be noted that even the schemes for implementing both actions are similar:

- the supplied energy is transferred to the potential energy inside the sample, due to the normal part of the external influence: in the case of welding, the explosives keep the missile plate on the fixed one; during torsion under pressure this function is partially performed by the anvil;
- the tangent part of the external influence largely forms the features occurring inside the sample: in the case of welding; it is the tangent part of the velocity of the flyer plate; for torsion under pressure, this is the angular velocity.

However, there are obvious differences: explosive welding is more rapid, the methods for applying energy to these methods are different.

In order to better link these two processes, we consider the article [116], which describes mechanical alloying by pressure torsion for an equiatomic Cu–Ta compound. A set of alternating plates (25 μm) of pure copper and tantalum was subjected to torsion under a pressure of 4 GPa for 10, 30, 50, 100, 150 turns, respectively. The thickness of the discs during rotation decreases from 0.90 mm to 0.70 mm. In this work, the effect of subsequent heat treatment on the phase composition and evolution of the microstructure of the compounds obtained was investigated.

Mechanical alloying is able to create alloys from elements that do not mix under equilibrium conditions. In work [116], the torsion under pressure was used to enhance the mechanical mixing of copper and tantalum, which do not mix at room temperature under the equilibrium conditions.

Figure 145 shows the SEM images of discs after torsion under pressure at 10 (*a*), 30 (*b*), 50 (*c*), 100 (*d*), 150 (*e*) revolutions. No bubbles or pores were detected by optical or electron microscopy, either in flat or in cross section. After 10 revolutions, the layers remain parallel (Fig. 145 *a*). After 30 revolutions, the layers are

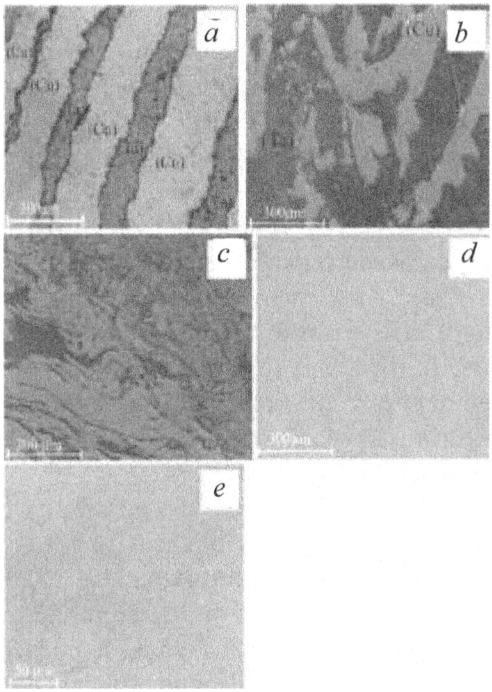

Fig. 145. SEM images of discs after torsion under pressure at 10 (*a*), 30 (*b*), 50 (*c*), 100 (*d*), 150 (*e*) revolutions [115].

destroyed and copper inclusions occur in the tantalum phase. As can be seen from (Fig. 145 *c*), even 50 revolutions are not enough for complete mixing, there remains a noticeable heterogeneity. After 100–150 revolutions, there is no difference in the contrast between copper and tantalum, which proves their intensive mixing.

Thus, the torsion under pressure of a set of pure Cu and Ta foils leads to the formation of a mixture of nanostructured phases enriched in Cu and Ta (Fig. 145 *e*, *f*). After 150 revolutions a dense solid nanostructured material is formed.

Comparison of the results for SPD torsion [116] with the results for explosive welding, presented in section 10.2.2, reveals a striking difference in microstructures that was difficult to expect. The package of initial plates turns into a multi-layer composite during explosive welding, but loses its lamellar essence and turns into a solid solution during SPD torsion with a large number of turns.

We believe that this is due to the difference in the duration of the methods. Time of SPD torsion is long enough for the disappearance of the plates. The explosion occurs almost instantaneously and the plates are preserved, cohesion occurs between them, the structure of the interfaces changes. It is these changes that we discussed above.

Conclusions to chapter 10

1. The results of the study obtained by explosive welding of multilayer composites based on steel, magnesium, as well as welded joints Nb–Cu, Ta–Cu.

2. Special attention is paid to intermediate layers, which play the role of buffers. It is shown that the introduction of a titanium intermediate layer between the functional layers of aluminium and magnesium, preventing the formation of intermetallic compounds, promotes the weldability of these materials.

3. For magnesium-based composites, the possibility of magnesium boiling and foaming is shown. Consequences to which the presence of foam can lead to during cooling are revealed.

4. The Nb–Ta and Ta–Cu welded joints contained the ultrafine-grained refractory (Nb or Ta) and Cu phases. For Ta–Cu welded joints, the presence of a non-equilibrium phase based on Ta or quasicrystals in the Cu matrix is also detected.

5. For multilayer Cu–Ta composites, it was first discovered that for the transition zone three states are realized simultaneously: a smooth surface, a flat boundary (consists of separate cusps) and a wave-like boundary. A similar surface topography for insulated welded joints was not observed.

6. The package of initial plates turns into a multi-layer composite during explosive welding, but loses its lamellar essence and turns into a solid solution during SPD torsion with a large number of revolutions. We believe that this is due to the difference in the duration of the methods.

11

Self-organization processes

In this chapter, we will analyze the main mechanisms of self-organization that occur during explosive welding of contacting materials near the interface and thereby ensure adhesion of these materials [82, 83, 89]. When welding is carried out, a number of problems are solved from the very beginning, including the choice of starting materials, the preparation of the surface of the plates for welding, the choice of explosives, the geometry of the explosive impact, etc., which in general allows a certain control over the process of producing a welded joint. In this case, a very significant role is played by the physical characteristics and behaviour of the system of welded plates, to which a large amount of energy is supplied during the explosion. The main part of it, due to the chemical energy of explosives, is the kinetic energy of the flyer plate. The formation of a strong welded joint is most likely possible if the kinetic energy is converted into a change in the internal energy of materials localized near the contact surface. Such a transformation can occur due to various dissipative processes, namely, those that have time to occur in a short time of explosive exposure. These include the formation of different types of cusps on the interface, granulating fragmentation, the adiabatic transition of mechanical energy into thermal energy, melting processes, movement of some defects, etc. At the same time, it can be assumed that from all possible scenarios and methods for implementing this transformation such process is chosen for which the speed of the process, i.e. the rate of conversion of the input mechanical energy into internal energy will be maximum. In particular, the maximum possible should be, first of all, the rate of formation of new boundaries of contacting materials.

Next we will consider the processes leading to the formation of cusps and splashes on the interface, as well as to their further evolution and self-organization as the intensity of external influence increases. We confine ourselves to the fact that of the numerous microphotographs, here are the numbers of only a few typical ones.

11.1. Transitions from splashes to waves

For different joints, welding modes were used below the lower boundary (LB) of the 'window of weldability', near LB, above LB. The welding modes near the LB are characterized by minimal collision velocities, ensuring the formation of a strong joint [4].

Below the LB (Figs. 104–106), the cusps are sufficient for joining, but too small for weldability. On LB, the dominant process of self-organization is the formation of cusps such as splashes, which have a relatively regular distribution on the surface of the refractory phase (Figs. 82, 107, 120, 122). Some of them form groups of splashes pressed to each other, elongated along the selected direction. This direction is parallel to the line of intersection of the planes of the falling and fixed plates. With further intensification of the welding mode, slightly higher than LB, instead of a simple option — increasing the height of the splashes — the system goes another way: aligning groups of cusps into rows along a selected direction, increasing groups of cusps both in height and length, and increasing the distance between rows and the subsequent transformation of groups into waves.

In this case, such a factor as the mutual solubility of the starting materials becomes significant. In the absence of mutual solubility, for example, for copper–tantalum welded joints, such large groups of splashes are observed when the line between them and the waves is erased (Fig. 108 *d*). Starting from a certain moment, when a complete docking of the cusps occurs, the mechanism of the zip-fastener is activated and only waves remain. However, instead of the formation of a perfect wavy interface, the system, as the external influence increases, passes through a series of intermediate states. They may have a 'patchwork quilt' type structure consisting of disoriented regions filled with waves with different parameters and splashes between regions (Fig. 107, *b, d*, 108, *f*).

With limited solubility of the starting materials, for example, for copper–titanium or aluminium–tantalum welded joints, the principle remains the same, but the sequence of structures may also include other elements. In this case, the surface of the cusps is covered with intermetallic particles. Such particles, in particular, interfere with the action of the zip-fastener mechanism mentioned above just as for a real zip-fastener, for example, the presence of dust would interfere. As a result, in the early stages of its formation, the wave is intermittent (Figs. 116, 117): at a certain length, it is a dense packing of cusps, then breaks off, then again cusps closely spaced to each other, etc., arise. Quasi-wave structures of the type of 'quilt' (Fig. 118 *c*, 120 *b*) are also formed from intermittent waves, but of a different type than those considered above. Figure 119 shows photographs of this patchwork, the analogy with which is used to describe the quasi-wave surface observed during explosive welding.

Here, to some extent, we describe the behaviour of the system under study, by analogy with the behaviour of a sequence of successive biological populations whose self-organization is aimed at their survival and development. For the system under study, these are splashes, large groups of splashes and quasi-wave structures of the 'quilt' type.

It is possible to doubt whether it is legitimate to use such common notions as 'wedges', 'splashes', 'patchwork quilt' to describe the behavior of two materials that have experienced a strong external effect. In our opinion, this is partly justified by the fact that it is precisely such concepts that make it possible to visually describe the observed picture.

The reasons why transitional structures were not previously observed were as follows. Usually, only the wavy surface of the section is examined due to its greater practical application and the flat surface is not examined. As a result, the area near LB or below LB is not used. But then you can not expect the observation of any splashes or rare cusps that provide a setting. In addition, a wave-like surface is usually obtained under modes inside the weldability window. At the same time, such an intermediate structure as a 'patchwork quilt' is lost, since the area that is higher than LB is not used, but only slightly. To observe these structures, it is also necessary to be able to completely etch one material and observe the surface of another. This contributes to the high corrosion resistance of tantalum and titanium.

At a distance from the lower boundary of the 'weldability window', a wavy interface forms, one of the most characteristic and striking phenomena that distinguishes the process of explosive welding from other methods of joining materials. Numerous attempts have been made to explain the mechanism of wave formation.

One of the first attempts to explain the nature of wave formation belongs to Abrahamson [117]. In work [117] special attention is paid to the experimental fact of the appearance of waves at a given velocity of collision of a steel bullet with a thin lead target at a collision angle greater than a certain critical one. Figures 146 *a, b* are images of the end surface of a steel bullet after a collision with a lead target at different values of the angle of impact, and in Fig. 146 *c* – images of the surface of the steel plate after the collision with a copper plate, thrown by means of an explosion. Such a relief looks like a 'patchwork' of a strip type (Fig. 118 *c,* 120 *b*).

In addition, another experiment simulating wave formation was carried out in [117]: a stream of water, remaining stationary, fell obliquely onto a viscous moving bottom, slowly moving. As a result, it was possible to obtain a periodic deformation of the surface.

In [118...120] hydrodynamic models are also used. However, the nature of wave formation is still not fully understood and there is no generally accepted theory of this phenomenon. However, the very emergence of a periodic surface topography remains surprising.

This section also attempts to explain the mechanism of wave formation during explosive welding. It takes into account the fact that, due to the strong impact of a plate collision, there is a significant deflection of materials, as a result of which the contact area increases significantly. Further, after the end of the external impact, the material as a whole returns to its original state, and the excess contact surface should, on the one hand, remain, and on the other, return to its original state. This relaxation can cause the wave-like nature of the interface.

11.2. Simulation experiments

In order to clarify the question of the possible ways of relaxation of a non-equilibrium structure with an excess area, a series of simulation experiments was conducted for a number of contacting materials [121]. Residual effects occurring after the bending deformation of two adjoining materials were investigated.

A simple scheme was used: a metal plate was attached to the ruler at two fixed points. A steel or plastic ruler was used. The plate was made of aluminium, copper, tantalum or lead. The thickness of the ruler is an order of magnitude greater than the thickness of the

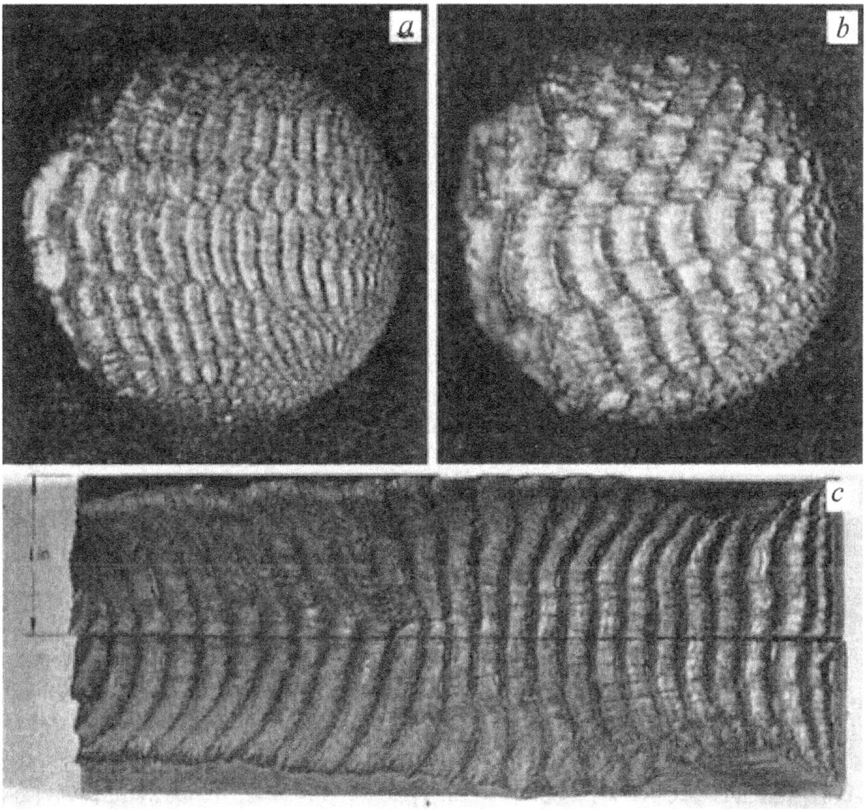

Fig. 146. The wavy surface of the steel as a result of a strong external impact: *a*, *b* – collision of a steel bullet with a lead plate (different angles of collision); *c* – collision of the flyer copper plate and steel plate [117].

plate. Therefore, in the case when the ruler is experiencing elastic deformation, the plate would experience plastic deformation.

In the case when the plate was not glued, it all, except for fixed points, experienced a deflection, which was maintained when the load was removed. As a result, the disc took the form similar to the dome. A completely different form was the plate, which was originally glued to the ruler. As can be seen from Figs. 147...148, the excess area goes to the formation of several kinks of the 'roof' type. It should be emphasized that the relaxation considered here, consisting in the formation of kinks, occurs at room temperatures and for sufficiently long times.

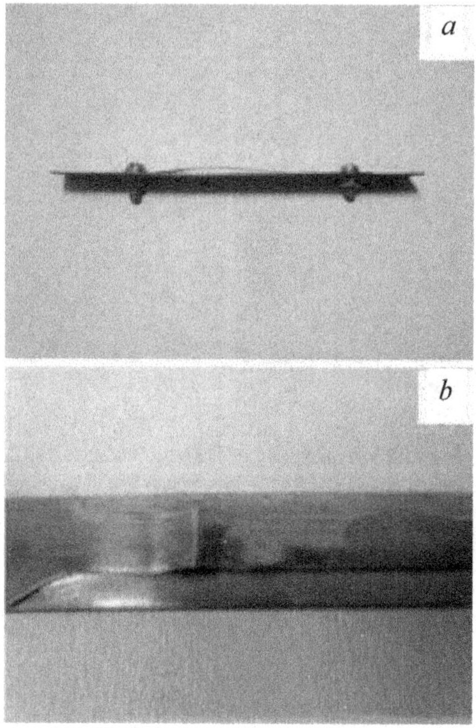

Fig. 147. Steel ruler – tantalum plate: *a* – deflection of a 'dome' type plate (the plate is not glued); *b* – deflection of the 'roof' plate (the plate is glued).

Here we present in somewhat more detail the results of a simulation experiment for a pair formed by a steel ruler and a tantalum plate. After the deflection, the ruler always restored the flat shape, whereas the plate, not glued to the ruler, really took the form of a dome (Fig. 147 *a*). But if the plate was glued to the ruler, then there appeared a kink similar to the 'roof' (Fig. 147 b). For this pair, a video was also taken, which can be found in [122]. The sequence of frames corresponds to the scenario: the ruler together with the tantalum plate glued to it strongly bends, and then gradually released and at some point a kink appears on the plate.

Figure 148 shows the results of a simulation experiment for pairs of steel ruler–glued plate of different metals: aluminium, copper, lead. Found that in all cases there are kinks, i.e. plate deformation is irreversible.

Further, the experiment was expanded to find out the difference in the behaviour of the plate in the case when it is glued to either

Fig. 148. The deflection of plates of aluminium, copper, lead, glued to the steel ruler.

the stretchable side of the ruler, as in previous cases, or to the compressible one. The ruler was made of plastic, and the plate was made of lead. The thickness of the ruler is 2 mm, the plate thickness is 0.1 mm. The radius of curvature was 30 mm for samples A and 20 mm for samples B. In Fig. 149 *a, b* the plate deflection is shown for samples A and B, respectively, for the case when there was no gluing, For sample A, the dome mentioned above is observed; for sample B, the deflection had a somewhat more complex shape, possibly due to the larger value of excess length.

As can be seen from Figs. 149 *c–g*, kinks occur only if the plate is glued to the stretchable side of the ruler. However, if the plate is glued to the compressible side of the ruler, then in this case it still does not remain flat – its surface becomes wrinkled. Figures 149 *c* and *e* shows for sample A the shape of the plate glued to the stretchable (*c*) and compressible (*e*) sides of the ruler. Figure 149 *d* shows the form of the plate, glued to the so-called midline [123]. As expected, the plate in this case remains flat, since there are no stresses on the middle line. For a sample with a radius of curvature smaller than for sample B, we obtain the greatest number of kinks,

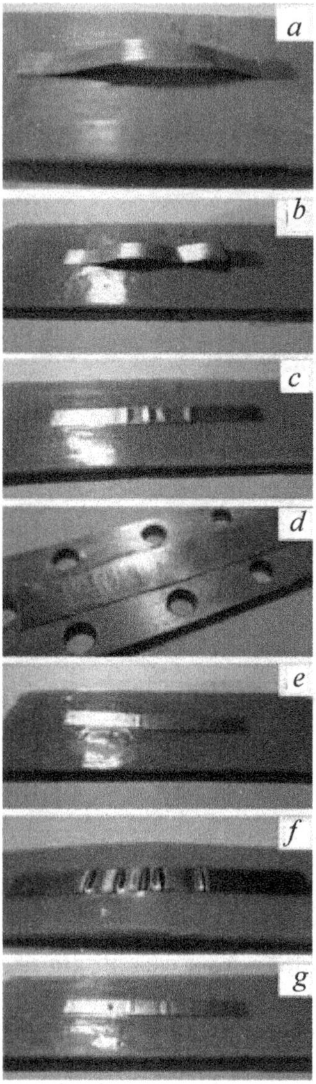

Fig. 149. Plastic ruler – lead plates: *a*...*b* – plate deflection (not glued); *c*...*g* – deflection of a plate glued to a stretchable or compressible side of the ruler, with different radii of curvature.

which, as can be seen from Fig. 149 *f*, is equal to five. On the inner side of the bend (Fig. 149 *g*), the relief change still remains weak.

Figures 147...149 show that the 'roof' is hollow, i.e. kinks occur due to the separation of the test plate from the carrier line. This is due to the fact that the contact surface of the ruler and the plate is inhomogeneous and in the weakest places there is a break and the formation of a break.

Imitation experiments, the results of which are presented here, have only a remote connection with explosive welding. In this regard, they are similar to the Abrahamson experiments [117] mentioned above, in which the occurrence of a periodic relief, either in the case of a bullet colliding with a target, or in the case of a moving surface under the action of a falling jet, simulates the possibility of wave formation during explosive welding. Similarly, the occurrence of inhomogeneities of the contact surface as a result of the realization of excess area also simulates wave formation. This is the significance of simulation experiments, although they do not provide sound models.

Conclusions to chapter 11

1. In the transition from a flat interface to a wave-like surface, a quasi-wave surface is observed, consisting of disoriented regions with their own wave parameters. In its structure, the quasi-wave surface is similar to a 'quilt', the shape of which depends on the type of mutual solubility of the original elements.
2. The similarity of the relief of the surface of the steel bullet at collision with a target, observed in the well-known Abrahamson experiments [117], and observed during explosive welding for a quasi-wave surface of the 'patchwork quilt' type, having a band structure.
3. Simulation experiments were carried out in order to find out possible ways of relaxation of a non-equilibrium structure with an excess area. At the same time, residual effects after bending deformation of two adjoining materials were investigated. Under certain conditions, it has been found that fractures occur on a plate made of various metals (lead, aluminium, copper, tantalum) imitating irregularities at the interface when welding by explosion

Conclusion

In this monograph, the structure of several dozen welded joints is investigated. The welding parameters – impact angle and contact velocity – are shown in Figs. 1, 103, 113 and 130.

The reader can pay attention to a rather modest citation volume. Indeed, the number of links is less than the number of pictures. We, for our part, pay attention to the variety of topics and materials that are affected, apart from explosive welding, when quoting: plastic deformation and fracture, high pressure shear (SPD by torsion), self-organization, dissipative channels, vortex formation, colloidal solutions, fractals, chemical reactor, petrochemical reactor (coke oven), metals, intermetallic compounds, composites, semiconductors, ceramics, glass.

The presented results of electron microscopic study of the structure may seem to some extent redundant. However, they allowed us to identify the processes that control the formation of joints during explosion welding. As we wrote in the Introduction, this turned out to be not at all hopeless. There are really not many such processes. The main ones are the fragmentation of two types, the formation of cusps, splashes and waves, local melting [124, 125].

Local melting is observed for many welded joints. For the first time, we paid attention to the occurrence of colloidal solutions during explosive welding and formulated the necessary relations between characteristic temperatures. For the copper–tantalum welded joints, the local melting zone is filled with a suspension of tantalum in copper (melted and then frozen), which provides dispersion hardening. In addition, these zones provide point gluing of the metals being welded. They form numerous ‘capsules’ randomly distributed over the interface between the tantalum and copper layers in the wall of the chemical reactor. And the presence of such ‘capsules’ does not interfere with the stability of the structure and the preservation of the unique corrosion resistance of tantalum for a long time of operation. This is, in our opinion, convincing evidence that local melting is not always dangerous.

Fragmentation of a single type is similar to fragmentation under severe deformation, observed in many materials. We were able to detect and identify a new, previously unknown, process of fragmentation, which owes its existence to an explosion [126].

When discovering a new phenomenon, it is necessary to give it a name so that it can be integrated into a circle of already known phenomena or structures. Fragmentation due to explosion welding was called granulating fragmentation (GFR). It is this type of fragmentation that is closest to explosion fragmentation.

Granulating fragmentation is a universal phenomenon: there was not one of the studied welded joints wherever it was found. This is the main dissipative channel through which the energy supplied by the explosion goes. Moreover, it was found that for ceramics and glasses, the transformation of a powder into a solid material during SPD torsion occurs as a result of the GFR process, which includes crushing of particles and their consolidation. At the same time, it is obvious that the GFR processes for various strong effects are not identical. As can be seen from the numerous above figures, the plate that appears after the SPD consists of microvolumes separated by microcracks. In this case, the grains are not observed due to the absence of dislocations. In contrast to the GFR, when welding by explosion, in the case of SPD by torsion, particle scattering and melting are also not observed. We believe that in addition to fragmentation and consolidation, the formation of the plate is based on the effect known as 'glass sticking'. In this case, it is valid for each of the studied materials, not only for glasses.

Irregularities were found on the interface, which were called cusps. They act like 'wedges', connecting the contacting surfaces with each other. It was found that under certain conditions the cusps resemble splashes on water, although they are solid. The self-similar nature of splashes during explosion welding was a factor that initiated the fractal description of the interface. We believe that splashes are precursors of waves. And then their role increases dramatically, since the problem of wave formation is fundamental and not yet solved.

Two forms of welded joint interface are known: flat and wavy. The study of transition states between them has not previously been conducted. It requires a set of joints produced by intermediate welding conditions. As a result, a sequence of structural states of the interface was identified, replacing each other as the welding mode intensified. Below the lower boundary (LB) of the weldability window, the structure observed in the setting area consists of isolated cusps. Splashes appear on a flat surface near LB. Slightly higher than LB, there is a structure called a 'patchwork quilt', which is a piecewise wave surface containing both waves and splashes. Inside the weldability window, closer to its centre, a fairly perfect wavy surface is observed.

We have already paid attention to the analogy between the surface observed during explosion welding and the real surface containing splashes. This analogy is deeper and extends to different types of waves, both stationary and non-stationary. Wave photographs from the Rey Collins [127] collection are shown in Fig. 150. Images of real waves taken by a famous photo artist are impressive. We draw attention to the similarity of images of real waves and waves during welding, shown in the inset to the corresponding figures.

One of the most successful implementations of explosive welding was used to obtain the wall of a chemical reactor. The reason for its high quality and stability is the stationary nature of the spatially homogeneous wavy interface surface of copper-tantalum. If for the composite forming the wall, the surface would be in a transition state, then during operation such a state would change, moreover, due

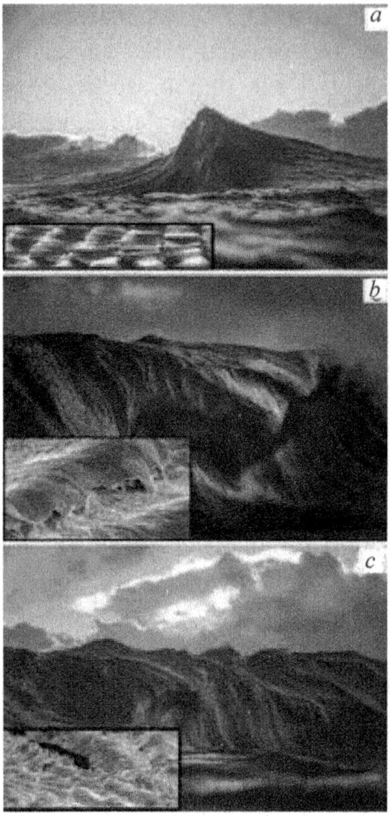

Fig. 150. Waves of various shapes (from the Rey Collins blog "Zamorozhennyie-morskie-volnyi" (Frozen sea waves) [127]); on the side: *a* – Fig. 110, *a; b* – Fig. 108, *b; c* - Fig. 108 *d.*

to the heterogeneity of deformation in different places differently. As a result, the walls of the reactor could not withstand long periods of operation. Other reasons were discussed above in chapter 5.

Using the well-known expression "Nomen est omen", which is translated as "Name is a sign," we give the names of the processes and structures discovered by the authors of the monograph:

- granulating fragmentation,
- cusps,
- splashes,
- quasi-wave interface.

The results presented in the monograph may give our readers information for thought, on the basis of which they will propose new ideas, possibly different from those developed by the authors.

Welding was carried out at the Volgograd State Technical University, FSUE Central Research Institute of Prometey, and

Uralkhimmash OJSC. Structural studies were carried out at the Center for collective use of electron microscopy of the Ural Branch of the Russian Academy of Sciences, at the Volgograd State Technical University and Belgorod State University.

References

1. Deribas, A.A. Physics hardening and explosion welding., A.A. Deribas. - Novosibirsk: Nauka, 1980. –220 p.
2. Lancaster J.F., Metallurgy of welding, J.F. Lancaster , (6^{th}ed), Abington Cambridge: Abington Publishing. 1999. - P. 446
3. Lysak, V.I., Explosion welding..Moscow. Mashinostroenie. 2005.
4. Lysak V.I., Lower boundary in metal explosive welding. Evolution of ideas, V.I. Lysak, S.V. Kuzmin , Journal of Materials Processing Technology. 2012. - No. 212. –- P. 150–156.
5. Konon, Yu.A. Explosion welding.,Yu.A. Konon, L.B. Pervukhin, A.D. Chudnovsky. - Moscow: Mashinostroenie, 1987.
6. Ryabov, V.R. Welding dissimilar metals and alloys.,V.R. Ryabov, D.M. Rabkin, R.S. Kurochko. - Moscow Mashinostroenie, 1984. –239 p.
7. Zakharenko, I.D. Welding metal explosion.,I.D. Zakharenko. - Minsk: Nauka and Tekhnika, 1990. –205 p.
8. Rybin, V.V. Bimetallic compound of orthorhombic titanium aluminide with titanium alloy (diffusion welding, explosion welding), V.V. Rybin, V.A. Semenov, [and others] , Materials Science. 2009. - V. 59. - No 3. - p. 372–386.
9. Rybin V.V. Examining the Bimetallic Joint Orthorhombic Titanium Aluminide and Alloy (Diffusion Welding), V.V. Rybin, B.A., Greenberg, O.V. Antonova, L.E. Kar'kina, A.V. Inozemtsev, V.A. Semenov, A.M. Patselov, Welding Journal. 2007. - V.86. - No7. - P. 205-s – 210-s.
10. Greenberg B.A. (Cu – Ta, Fe – Ag, Al – Ta), B.A. Greenberg, M.A. Ivanov, V.V. Rybin, O.A. Elkina, O.V. Antonova, A.M. Patselov, A.V. Inozemtsev, A.V. Plotnikov, A.Yu. Volkova, Yu.P. Besshaposhnikov, Materials Characterization. 2013. - V. 75. - P. 51–62.
11. Greenberg, B.A. Heterogeneity of the interface during explosion welding, B.A. Greenberg, M.A. Ivanov, V.V. Rybin [et al.] , Fiz. Met. Metalloved. - 2012. - V. 113 - No 2. - p. 187–200.
12. Greenberg B.A. Fragmentation processes during explosion welding (Review), V.A. Greenberg, M.A. Ivanov, V.V. Rybin, O.A. Elkina, A.M. Patselov, O.V. Antonova, A.V. Inozemtsev, T.P. Tolmachev , Russian Metallurgy (Metally). 2013. - V. 10 No, - P. 727.
13. Greenberg B.A. Dissipative structures during explosion welding, B.A. Greenberg, M.A. Ivanov, V.V. Rybin [et al.] , Series Explosion Welding and Properties of Welding Connections., Izv. VolgGTU. - Vol. 5. - No. 14. - p. 27–43.
14. Greenberg B.A. The Processes of Fragmentation, Intermixing and Fusion upon Explosion Welding, B.A. Greenberg, M.A. Ivanov, A.M. Patselov, Yu.P. Besshaposh-

nikov , AASRI Procedia (WIC 2012). V. 3. 2012. - P. 66–72.
15. Banerjee D. The Physical Metallurgy of Ti3 Al Based Alloys, D.Banerjee, A.K. Gogia, T.K. Nandy, et al., Structural Intermetallics. ed. R. Darolia et al. - Warrendale. PS: AIME. 1993. P. 19–31.
16. Ward C.H. Microstructure of Ti-Al-Nb $\alpha_2$intermetallics, C.H. Ward , Intern. Mater. Rev. 1993. V. 38. No2. - P. 79–101.
17. Mirasl D.V. Phase Equilibria in Ti-Al-Nb, D.V. Mirasl, M.A. Foster C.G. Rhodes, Proceedings of the Titanium'95 conference: Science and Technology, 1996. - P. 372–379.
18. Flower H.W. Phase equilibria and transformations in titanium aluminides, H.W. Flower, J. Christodoulou , Mat. Science and Technology. 1999. V. 15 (1). - p.45–52.
19. Rhodes C.G. Order, disorder temperature of the bcc phase in You 21Al- 26Nb, C.G. Rhodes , Scr. Mater. 1998. V. 38 (4). - p. 681–685.
20. Greenberg B.A. Anomalies of Deformation Behavior of TiAl intermetallic, B.A. Greenberg, M.A. Ivanov , Advances in Metal Physics. 2000. 1 - No1. - p. 9–48.
21. Greenberg, B.A. Ni_3Al and TiAl intermetallic compounds: microstructure, deformation behavior, B.A. Greenberg, M.A. Ivanov, Ekaterinburg: Ural Branch of the Russian Academy of Sciences, 2002. –360 p.
22. Greenberg B.A. Microstructure of bimetallic joint of titanium and orthorhombic titanium aluminide (explosion welding), B.A. Greenberg, V.V. Rybin, O.V. Antonova, In: monograph Severe Plastic Deformation: Tovard Bulk Production of Nanostructured Materials, New-York: Nova Scene Publishers Inc. 2005, p. 533-544.
23. Rybin, V.V. Structure of the transition zone in explosion welding (titanium - orthorhombic titanium aluminide), V.V. Rybin, B.A. Greenberg, M.A. Ivanov, O.V. Antonova, O.A. Elkina, A.V. Inozemtsev, A.M. Patselov , Svarka i diagnostika. 2010. - No 3. - p. 26–31.
24. Greenberg, B.A. The processes of melting, vortex formation and fragmentation during explosion welding, B.A. Greenberg, M.A. Ivanov, V.V. Rybin [et al.] , Svarka i diagnostika. 2010. - No 6. - P. 34–38.
25. Rybin V.V. Structure of the Welding Zone between Titanium and Orthorhombic Titanium Aluminide for Explosion Welding: I. Interface, V.V. Rybin, V.A. Greenberg, M.A. Ivanov, S.V. Kuz'min, V.I. Lysak, O.A. Elkina, A.M. Patselov, A.V. Inozemtsev, O.V. Antonova, and V.E. Kozhevnikov , Russian Metallurgy (Metally). 2011. - V. 10 No, - R. 1008.
26. Greenberg B.A. Orthorhombic Titanium Aluminide for Explosion Welding: II. Local Melting Zones, B.A. Greenberg, M.A. Ivanov, V.V. Rybin, S.V. Kuz'min, V.I. Lysak, O.A. Elkina, A.M. Patselov, O.V. Antonova, and A.V. Inozemtsev , Russian Metallurgy (Metally) 2012. - V.10 No, - R.1016.
27. Anoshkin, N.F. (Ed.) Metallography of titanium alloys., N.F. Anoshkin. - Moscow, Metallurgiya, 1980. –464 p.
28. Kolachev, B.A. Titanium alloys of different countries, B.A. Kolachev, I.S. Polkin, V.D. Talalaev. - Moscow VILS, 2000. –316 S.
29. Zwicker, U. Titan and its alloys., W. Zwicker - Moscow, Metallurgiya., 1979. –511 p.
30. Salishchev, G.A. Use of superplastic modes deformation for the manufacture of products from intermetallic compounds, G.A. Salishchev, R.M. Imaev, A.V. Kuznetsov, etc., Kuznechno-shtamp. proizvodstvo.1999. - No 4. - p. 23-28.
31. Bondar, M.P. The influence of the deformation mechanism in the area of collision of pairs of materials on the choice of the optimal parameters of explosion welding, M.P. Cooper , Avt. Svarka. 2009. - No 11. - P. 14–18.

32. Cahn, R. Physical metallurgy., R. Cahn , Vol. 2. - Moscow Mir, 1968. –490 p.
33. Feder, E. Fractals, E. Feder. - Moscow, Mir, 1991. 254 p.
34. Greenberg, B.A. The structure of the transition zone in explosion welding (copper - tantalum), B.A. Greenberg, M.A. Ivanov, V.V. Rybin [et al.] , Deform. Razrush. Mater.2011. - No 9. - p. 34–40.
35. Greenberg B.A. Cu-Ta joint made by explosion welding, V.A. Greenberg, O.A. Elkina, O.V. Antonova, A.V. Inozemtsev, MA Ivanov, V.V. Rybin and V.E. Kozhevnikov, The Paton Welding Journal. 2011. - V. 7. - P. 20–25.
36. Greenberg, B.A. Processes and structures in explosion welding, B.A. Greenberg, O.A. Elkina, A.M. Patselov [et al.] , Svarka i diagnostika. 2013. - No2. - p. 35–40.
37. Frey D. Recent Successes in Tantalum Clad Pressure Vessel Manufacture: A New Generation of Tantalum Clad Vessels, D. Frey, J. Banker, Retress Proceedings of Corrosion Solutions Conference 2003. USA: Wah Chang, 2003. - P. 163 –169.
38. Greenberg V.A. Problems of stirring and melting in explosion velding (aluminium — tantalum), V.A. Greenberg, O.A. Elkina, A.M. Patselov, A.V. Inozemtsev, A.V. Plotnikov, A.Yu. Volkova, M.A. Ivanov, V.V. Rybin and Yu.P. Besshaposhnikov , The Paton Welding Journal. 2012. - V. 7. - R. 12–19.
39. Volkova A.Yu. Electron-Microscopic Examination of the Transition Zone of Aluminium-Tantalum Bimetallic Joints (Explosion Welding), A.Yu. Volkova, V.A. Greenberg, M.A. Ivanov, O.A. Elkina, A.V. Inozemtsev, A.V. Plotnikov, A.M. Patselov, and V.E. Kozhevnikov , Physics of Metals and Metallography. 2014. - V.115. No, pp. 380–391.
40. Greenberg B.A. Explosive Welding: Mixing of Metals without Mutual Solubility (Iron-Silver), V.A. Greenberg, M.A. Ivanov, V.V. Rybin, O.A. Elkina, A.V. Inozemtsev, A.Yu. Volkova, S.V. Kuz'min, and V.I. Lysak , Physics of Metals and Metallography. 2012. - V. 113. - No 11. - P. 1041-1051.
41. Greenberg, B.A. Identification of risk zones for the shell of a petrochemical reactor (coke oven chamber) obtained by explosion welding, B.A. Greenberg, O.A. Elkina, A.M. Patselov [et al.] , Svarka i diagnostika. 2014. - No3. - p. 4–9.
42. Mott N.F. Fragmentation of Shell Cases, N.F. Mott, Proc. Royal Soc. (January, 1947). A189. - P. 300–308.
43. Grady D. Fragmentation of the Rings and Shells: The Legacy of N.F., D. Grady , Mott. Springer-Verlag Berlin Heidelberg, 2006. - P. 361.
44. Orlenko, L.P. Physics of the explosion., L.P. Orlenko. - Moscow FIZMATLIT, T.2. 2002. –832 p.
45. Vladimirov, V.I. The physical nature of the destruction of metals., - Moscow, Metallurgiya., 1984. –280 p.
46. Rybin, V.V. Crossing of grain boundaries by slip bands as a mechanism of viscous grain-boundary fracture, V.V. Rybin, V.A. Likhachev, A.N. Vergazov, Fiz. Met. Metalloved. 1973. - V. 36. - No5. - p. 1071-1078.
47. Rybin, V.V. Large plastic deformation and destruction of metals., V.V. Rybin - Moscow, Metallurgiya., 1986. –224 p.
48. Zhilyaev A.P. Using high pressure torsion for metal processing: Fundamentals and applications, A.P. Zhilyaev, T.C. Langdon, Progress in Materials Science V.53. - 2008. - P. 893–979.
49. Valiev, R.Z. Bulk nanostructured metallic materials, R.Z. Valiev, I.V. Alexandrov - Moscow: Akademkniga, 2007. –398 p.
50. Kolobov Yu.R. Grain boundary diffusion and properties nanostructured materials., Yu.R. Kolobov, R.Z. Valiev, G.P. Grabovetskaya [et al.] - Novosibirsk: Nauka, 2001. –232 c.

51. Rabier J. Dislocations and plasticity in semiconductors. I - Dislocation structures and dynamics, J. Rabier, A. George , Rev. de Phys. App. 1987. V. 22. - No 9. - p. 941–966.
52. Islamgaliev R.K. Structure of silicon processed by severe plastic deformation, R.K. Islamgaliev, R. Kuzel, S.N. Mikov et al./Mat. Sci. Eng. 1999. V. A266. - P. 205–210.
53. Greenberg, B.A. The role of fragmentation of the type of crushing in the consolidation of powders of quartz ceramics and glass under torsion under pressure, B.A. Greenberg, M.A. Ivanov, V.P. Pilyugin [et al.] , Deform. Razrush. Mater. 2016. - No12. - p. 17-26.
54. Stremel, M.A. Destruction. Destruction of materials, M.A.Shtremel - Moscow Izd. House MISiS, Vol.. 1. 2014. – 670 s.
55. Stremel M.A. Destruction Destruction of materials, M.A. Shtremel - Moscow Izd. House MISiS, Vol. 2. 2015. –976 s.
56. Ailer, R. Chemistry of silica., R. Ailer , Moscow Mir, 1982. - Part 1. - 416 p.
57. Plotnikov, A. V. Is it possible to automatically block dislocations after torsion under pressure?, A.V. Plotnikov, B. A. Greenberg, M. A. Ivanov, V. P. Pilyugin, T. P. Tolmachyov, O. V. Antonova, M. A. Patselov, Physics of Metals and Metal Science, 2017. - V. 118 - p. 843–849.
58. Greenberg A.B. Microheterogeneous Structure of Local Melted Zones in the Process of Explosive Welding, A.V. Greenberg, M.A. Ivanov, A.V. Inozemtsev, A.M. Patselov, M.S. Pushkin and A.M. Vlasova, Metallurgical and Materials Transactions A, 2015. - V. 46. - No 8. - p. 3569–3580.
59. Czerwinski F. Magnesium alloys-design, processing and properties (Edited by Frank Czerwinski), InTech, 2011. - R. 526.
60. Rybin V.V. Formation of Vortices during Explosion Welding (Titanium - Orthorhombic Titanium Aluminide), V.V. Rybin, B.A. Greenberg, O.V. Antonova, O.A. Elkina, M.A. Ivanov, A.V. Inozemtsev, A.M. Patselov, I.I. Sidorov , Physics of Metals and Metallography. 2009. - V.108. - No 4 - R. 353–364.
61. Rybin V.V. Nanostructure of Vortex During Explosion Welding, V.V. Rybin, V.A. Greenber, M.A. Ivanov, A.M. Patselov, O.V. Antonova, O.A. Elkina, A.V. Inozemtsev, G.A. Salishchev, Journal of Nanoscience and Nanotechnology. 2011. - V.11. - No10. - p. 8885–8895.
62. Frost, W. Turbulence: Principles and Applications, W. Frost, T. Moulden. - M: World ", 1980. - 535 p.
63. Monin, A.S. Statistical hydromechanics. Part 1. Mechanics of turbulence., A.S. Monin, A.S. Yaglom - Moscow. Nauka, 1965. - p. 639.
64. Landau, L.D., Theoretical physics. T. VI. Hydrodynamics, L.D. Landau, E.M. Lifshits- Moscow. Nauka, 1986. - p. 736.
65. Godovikov, A.A. Agates, A.A. Godovikov, O.I. Ripinen, S.G. Motorin - Moscow Nedra, 1987. –368 p.
66. Summ, B.D. Basics of Colloid Chemistry, BD Summ - Moscow Academy, 2009. –240 p.
67. Freydin, A.S. Properties and calculation of adhesive compounds, A.S. Freydin, R.A. Turusov - Moscow Khimiya, 1990. –256 p.
68. Pocius, A.V. Adhesives, adhesion, bonding technology, A.V. Pocius. - St. Petersburg: Professiya, 2007. –376 p.
69. Plotnikov, A.V. Composite structure for the wall of a coke oven chamber produced by explosion welding, A.V. Plotnikov, B.A. Greenberg, O.A. Elkin [et al.] , Oil. Gas. Novations. 2014. - No8. - p. 6–12.
70. Greenberg B.A. Risk zones for coke drum shell produced by eksplove welding, V.A.

Greenberg, O.A. Elkina, A.M. Patselov, A.V. Plotnikov, M.A. Ivanov, Yu.P. Bessha-poshnikov , Journal of Materials Processing Technologies 2015. - No15. - P. 79–86.

71. Ernst F. Enhanced Carbon Diffusion in Austenitic Stainless Steel Carburized at Low Temperature, F. Ernst, A. Avishai, H. Kahn, X. Gu, G.M. Michal, A.H. Heuer , Metall and Mat. Trans A. 2009. 40A. - p.1768-1780.
72. Cermak J. Carbon diffusion in carbon-supersaturated ferrite and austenite, J. Cermak, L. Kral , Journ. of All and Comp. 2014. V. 586. - P. 129–135.
73. Elliot, R. Managing eutectic solidification., R. Elliot [et al.] - Moscow, Metallurgiya., 1987. –352 p.
74. Houdremont E.A. Special steel. 2nd Edition., E.A. Gudremon - M.: Metallurgy, 1966. –734 p.
75. Andersson J.O. A Thermodynamic Evaluation of the Fe-Cr-C System, J.O. Andersson , Metall. Trans. A. 1988. - No 19 (3), - p. 627–636.
76. Computational Thermodynamics, Calculation of Phase Diagrams using the CALPHAD Method , Uddeholm AEB-L Stainless Steel, http://www.calphad.com/AEB-L.html
77. Bunin, K.P. Fundamentals of metal cast iron, KP. Bunin, J.N. Malinochka, Yu.N. Taran - Moscow, Metallurgiya., 1969. –415 p.
78. Greenberg B.A. Interface after Explosion Welding: Fractal Analysis, V.A. Greenberg, M.A. Ivanov, M.S. Pushkin, A.M. Patselov, A.Yu. Volkova, and A.V. Inozemtsev , Russian Metallurgy (Metally) 2015. - No10. - p .816.
79. Mandelbrot, B. The Fractal Geometry of Nature, B. Mandelbrot - Moscow: Institute for Computer Studies, 2002. - P. 656.
80. Lysak V.I. Structure of Boundaries in Composite Materials Obtained with Explosive Loading, V.I. Lysak, S.V. Kuz'min, A.V. Krokhalev, B.A. Greenberg , Physics of Metals and Metallography. 2013. - V. 114. - No 11. - R. 947.
81. Krokhalev, A.V. Thin structure of interphase boundaries in hard alloys of chromium-titanium carbide system, A.V. Krokhalev, V.O. Kharlamov, S.V. Kuzmin, V.I. Lysak, B.A. Greenberg , Proceedings of universities. Powder metallurgy and functional coatings. 2015. - No.2. - p. 38–43.
82. Greenberg B.A. Interface Relief upon Explosion Welding: Splashes and Waves, V.A. Greenberg, M.A. Ivanov, A.V. Inozemtsev, S.V. Kuz'min, V.I. Lysak, A.M. Vlasova and M.S. Pushkin , The Physics of Metals and Metallography, 2015. - V. 116. - No. 4. - p. 367–377.
83. Greenberg B.A. Evolution of Interface Relief during Explosive Welding: Transitions from Wavelets to Waves, V.A. Greenberg, M.A. Ivanov, A.V. Inozemtsev, S.V. Kuz'min, V.I. Lysak and M.S. Pushkin , Bulletin of the Russian Academy of Sciences. Physics, 2015. - Vol. 79. - No. 9. - R. 1118–1121.
84. Longuet-Huggins H.C. A computer algorithm for reconstructing a scene from two cusps, N.S. Longuet-huggins, Nature. 1981. V. 293. - No 10. - P. 133
85. Abramov, V.V. Reconstruction of three-dimensional surfaces in two cusps when the camera is tracking a given point of the scene., V.V. Abramov, V.S. Kirichuk, V.P. Kosykh, G.I. Peretyagin, S.A. Popov, Autometriya. 1998. - No 5. - p. 3–16.
86. Zhuk D.V. Restoration of a three-dimensional model of a scene from digital images, D.V. Zhuk, A.V. Tuzikov, A.V. Bearded, Iskusst. intellekt. 2006. - No 2. - p. 142–146.
87. Pushkin M.S. Detection of the quasi-wave shape of the interface when welding by explosion (copper-tantalum, copper-titanium), M.S. Pushkin, A.V. Inozemtsev, B.A. Greenberg [and others] , Abstracts of the report of the sixth international conference "Crystal physics and deformation behavior of promising materials", May 26 - 28,

2015, Moscow, p. 263.
88. Pushkin M.S. Wave shape of an interface (explosion – copper – tantalum, copper – titanium), M.S. Pushkin, A.V. Inozemtsev, B.A. Greenberg, A.M. Patselov, M.A. Ivanov, and O.V. Slautin , Bulletin of the Russian Academy of Sciences. Physics, 2016. - Vol. 80. - No. 10. - R. 1273–1278.
89. Greenberg, B.A. Self-organization and evolution of the interface in explosion welding (copper - tantalum, copper - titanium), B.A. Greenberg, M.A. Ivanov, A.V. Inozemtsev, [and others] , Fundamental problems of modern materials science. 2015. - No12. P. 391-402.
90. Diagrams of the state of double metallic systems: a Handbook in 3t., Under total ed. N.P. Lyakishev. - M: Mechanical Engineering, 1996. –992 p.
91. Greenberg, B.A. The structure of the melted zones in explosion welding (aluminium - tantalum, copper - titanium), B.A. Greenberg, M.S. Pushkin, A.M. Patselov, A.V. Inozemtsev, M.A. Ivanov, O.V. Slautin, Yu.P. Besshaposhnikov, Svar. Proiz. 2016. - - No5. - P. 25-35.
92. Gurevich, L.M. Intermetallic reactions in explosion welding (Cu-Ti), L.M. Gurevich, O.V. Slautin, M.S. Pushkin [and others], Izv. VolgGTU. 2015. - No10 (170). - pp. 32-37.
93. Gurevich, L.M. The formation of intermetallic compounds during explosion welding and subsequent heating, L.M. Gurevich, O.V. Slautin, M.S. Pushkin [et al. , Izv. VolgGTU. 2016. - V. 181. - P. 7-12.
94. Giessen B.C. A metastable phase TiCu3 (m), B.C. Giessen, D.Schimanski - J.Appl. Cryst. - 1971 - vol.4, - part 3 - p. 257-259.
95. Rievsk, J. Journal of Achievements in Materials and Manufacturing Engineering - 2012 - 50/1 - P. 26-39.
96. C.Z. Wagner , Electrochem V.65. 1961. - p. 581
97. Lifshitz I.M. Article title, I.M. Lifshitz, V. Slyozov , Journal of Phys. Chem. Solids. 1961. - No. 19. - p. 35
98. Cahn R.W. in Physical Metallurgy., R.W. Cahn , Elsevier, 1983. Ch. 25. Recovery and Recrystallization - P. 1596-1671.
99. Voevodin, L. B. To the question of the poor weldability of aluminium alloys , L. B. Voevodin , Metallography and strength of materials: interstitial. Sat scientific works, VolgPI. - Volgograd, 1983. - p. 83.
100. Kazak, N.N. Properties and the field of application of welded joints obtained by explosion welding. allowance, N.N. Kazak; Volgograd. ed. Volgograd Polytechnic. Inst., 1984, 77c.
101. Deribas, A. A. Determination of the limit modes of collisions that ensure the welding of metals by an explosion, A. A. Deribas, I. D. Zakharenko , Combustion and explosion physics. - 1975. T. 11, No1. - p. 151-153.
102. Relief of the interface when welding explosion of homogeneous materials, MS. Pushkin, A.V. Inozemtsev, B.A. Greenberg, A.M. Patselov, M.A. Ivanov, O.V. Slautin, Yu.P. Besshaposhnikov , Welding production. - 2017 - N7. - pp. 11-17.
103. Features of the structure of the interface for homogeneous materials obtained by explosion welding (copper-copper), A.V. Inozemtsev, M.S. Pushkin, B.A. Greenberg, M.A. Ivanov, O.V. Slautin, A.M. Patselov, Yu.P. Besshaposhnikov, Izv. VolgGTU. - 2017. - T.205, N10 - p. 26-31.
104. Bondar, M.P. Deformed state of the joint zone when welding with copper and copper by the explosion and the mechanism of its formation, M.P. Bondar, V.M. Ogolikhin, Sat. report 6th International Symposium on Explosive Energy, Gottwaldov. - 1985. - p. 338–345.

105. Pushkin, M.S. Inozemtsev A.V. Features of the evolution of the interface of homogeneous joints produced by explosion welding, MS. Pushkin, A.V. Inozemtsev , Abstracts of the XVIII All-Russian School of Seminar on the Problems of Condensed Matter Physics (SPFCS-2018), Yekaterinburg. - 2017. - p.132.
106. Fractal analysis of welded joints (Cu-Ta, Cu-Ti), B.A. Greenberg, M.S. Pushkin, A.P. Tankeev, A.V. Inozemtsev, Fundamental problems of modern materials science. - 2017. - V.14. - p. 445-452.
107. Maltseva, L.A. Metal Laminated Composites materials obtained by explosion welding: structure, properties, features of the structure of the transition zone, LA. Maltsev, D.S. Tyushlyaeva, T.V. Maltsev, M.V. Shepherds, N.N. Lozhkin, D.V. Inyakin, L.A. Marshuk , Deform. Razrush. Mater. - 2013. - N4. - p. 19–26.
108. Gladkovsky S.V. Formation of the mechanical properties and fracture resistances of sandwich composites based on the 09G2S steel and the Ep678, S.V. Gladkovsky, S.V. Kuteneva, I.S. Kamantsev, R.M. Galeev, D.A. Dvoynikov , Diagnostic, Resource and Mechanics of materials and structures - 2017. - V.6. - P. 71-90.
109. Ruifeng Liu Numerical study of Ti, Al, Mg plates on the interface behavior in explosive welding, L. Ruifeng, W. Wenxian, Z. Tingting, Y. Xiaodan , Science and Engineering of Composite Materials. - 2017; - V.24 (6). - P.833–843.
110. Vlasova A.M. Multilayer Mg – Ti Based Composites Produced to Explosion Welding: Risk Zones, A.M. Vlasova, V.A. Greenberg, S. V. Kuz'min, V. I. Lysak, Inorganic Materials: Applied Research. - 2016. - V. 3. - P. 402-408.
111. Diagrams of the state of metallic systems. Issue 31 (Ed. By L.A. Petrova) (Moscow: VINITI: 1987).
112. Gelfman M.I. Colloid chemistry, MI Gelfman, O.V.Kovalevich, V.P. Yustratov. - St. Petersburg. Lan', 2004.
113. PradeepK. Parchuri Benefits of the intermediate-layer formation at the interface of Nb, Cu and Ta, Cu explosive clads, P. Parchuri, S. Kotegawa, H. Yamamoto, K. Ito, A. Mori, K. Hokamoto , Materials & Design. - 2019. - V.166. - Article 107612.
114. Greenberg B.A. Formation of intermetallic compounds during welding, B.A. Greenberg, M.A. Ivanov, M.S. Pushkin, A.V. Inozemtsev, A.M. Patselov, A.P. Tankeyev, S.V. Kuzmin, V.I. Lysak , Metall. Mater. Trans. A. - 2016. - V.47A. - P. 5461-5473.
115. Besshaposhnikov Yu.P. About multilayer throwing plates moving detonation wave, Yu.P. Besshaposhnikov, V.V. Pai, A.A. Petunin, V.I. Chernukhin , Izv. VolgGTU - 2018. - N11, T.211. - pp. 22-27.
116. Ibrahim N. Mechanical alloying through high pressure torsion of tha immiscible Cu50Ta50 system, N. Ibrahim, M. Peterlechner, F. Emeis, M. Wegner, S.V. Divinski., Materials Science and Engineering A. - 2017. - V.685. - P. 19-30.
117. Abrahamson G.R. Permanent periodic surface deformation due to a traveling jet, G.R. Abrahamson , Journal of Applied Mechanics. 1961. V. 28. - No4. - P. 512-528.
118. Bahrani A.S. The mechanics of wave formation in explosive welding, A.S. Bahrani, T.J. Black, B. Crossland , Proceeding of the Royal Society, Series A, Mathematical and Physical Science. 1967. - V. 296. - No1445. - P. 123-136.
119. Hunt J.H. Wave formation in explosive welding, J.H. Hunt. the Philosophical Magazine. 1968. - V. 17. - No146. - p. 669-680.
120. Covan G. Flow configuration in colliding plates, G. Covan, A. Holtzman, Journal of Applied Physics. 1963. - V. 34. - No4. - P. 928-937.
121. Greenberg V.A. Wave Formation During Explosive Welding: the Relaxation of a Nonequilibrium Structure, V.A. Greenberg, M.A. Ivanov, S.V. Kuzmin, V.I. Lysak, M.S. Pushkin, A.V. Inozemtsev, A.M. Patselov, and A.V. Pasheev, The Physics of Metals and Metallography, 2016. - Vol. 117. - No. 12. - R. 1223–1229.

122. https://youtu.be/eI6gRlYKqgQ
123. Asaro R.J. Mechanics of Solids and Materials, R.J. Asaro, V. A. Lubarda, Cambridge University Press, 2006 - P. 866.
124. Greenberg B.A. Copper-Tantalum Joints and Their Role in the Construction of the Chemical Reactor (Explosive Welding), V.A. Greenberg, M.A. Ivanov, A.M. Patselov, A.V. Inozemtsev, M.S. Pushkin, S.V. Kuzmin, V.I. Lysak , Chapter in the book "Tantalum: Geochemistry, Production and Potential Applications", 2015. USA. New-York: Nova Science Publishers Inc. 2015. - P. 125-182.
125. Greenberg, B.A., Ivanov, M.A., Pushkin, M.S., Inozemtsev, A.V., Pacelov, A.M., Tankeyev, A.P., Kuzmin, S.V., and Lysak V.I. Formation of Intermetallic Compounds During Explosive Welding, Metallurgical and Materials Transactions A. 2016. - Vol. 47A. - P. 5461-5473.
126. Greenberg, B.A. Fragmentation during explosion and explosion welding, B.A. Greenberg, M.A. Ivanov , Works of the 11th International Conference "Modern Metallic Materials and Technologies" SMMT-15, June 23–27, 2015, St. Petersburg, - P. 273–292.
127. http://telegraf.com.ua/zhizn/muzhchinyi/1893400-zamorozhennyie-morskie-volnyi-ot-reya-kollinsa-foto.html

Index

L

M

O

Q

R

S

T

V

W